Insect Herbivore–Host Dynamics

Literature currently available on the population dynamics of insect herbivores tends to favour a top–down regulation of abundance, owing much to the action of natural enemies. This unique volume challenges this paradigm and argues that tree-dwelling species of aphids, through competition for resources, regulate their own abundance.

The biology of tree-dwelling aphids is examined, particularly their adaptation to the seasonal development of their host plants. When host-plant quality is favourable aphids, by telescoping generations, can achieve prodigious rates of increase which their natural enemies are unable to match.

Using analyses of long-term population censuses and results of experiments, this book introduces students and research workers to insect herbivore–host dynamics using the interaction between aphids and trees as a model.

A. F. G. DIXON is an Emeritus Professor in the School of Biological Sciences at the University of East Anglia. He has written over 200 papers on aphids and their natural enemies in scientific journals, and has written or edited 10 books. In 1992, he was awarded the Gregor Mendel Gold Medal by the Czech Academy of Science, in 2000 a medal of honour by Akademia Podlaska, Poland, and in 2001 became Laureate of the University of South Bohemia, Czech Republic.

Insect
Herbivore–Host
Dynamics

Tree-Dwelling Aphids

A. F. G. DIXON
University of East Anglia

CAMBRIDGE UNIVERSITY PRESS
Cambridge, New York, Melbourne, Madrid, Cape Town,
Singapore, São Paulo, Delhi, Tokyo, Mexico City

Cambridge University Press
The Edinburgh Building, Cambridge CB2 8RU, UK

Published in the United States of America by Cambridge University Press, New York

www.cambridge.org
Information on this title: www.cambridge.org/9781107402638

First published 2005
First paperback edition 2011

A catalogue record for this publication is available from the British Library

ISBN 978-0-521-80232-1 Hardback
ISBN 978-1-107-40263-8 Paperback

Contents

Contents

Preface

The objective of this book is to give students and research workers an introduction to herbivore–host dynamics using the interaction between aphids and trees as a model. The reason for this is that the biology and population dynamics of these aphids are well studied, and limiting the scope enables one to focus on the problem and not be distracted by the details of many different herbivore–host systems. I have tried to make the text comprehensible to undergraduate students by avoiding much of the ecological and entomological jargon that abounds in the literature.

As stated in the Introduction I am greatly indebted to past students for doing much of the work reported here and in particular to Pavel Kindlmann and Seamus Ward for help in developing the ideas. I wish to express special thanks to Simon Leather for allowing me access to his long-term census data on sycamore aphids, Professor Furuta for supplying the cover illustration, Diana Alden for preparing the figures and Jenny Gill for sorting out the many computing problems I encountered.

I dedicate the book to June, whose support made the task possible.

1

Introduction

The prodigious rate of increase of aphids has fascinated entomologists for centuries. Réaumur (1737), like Leeuwenhoek, thought aphids were hermaphrodite and calculated that one aphid may give rise to 5.9 billion over a period of six weeks. Bonnet (1745) was the first to appreciate that aphids were bisexual but could produce a succession of broods without males, a phenomenon that later became known as parthenogenesis (Owen, 1849). Huxley (1858) was also fascinated by parthenogenesis in aphids and calculated that after 10 generations, if they all survive, an aphid can give rise to a biomass equivalent in weight to 500 million stout men. Occasionally these extraordinary rates of increase are realized. White (1887), of Selborne, records that at about 3 p.m. on 1 August 1774, showers of hop aphids fell from the sky and covered people walking in the streets and blackened vegetation where they alighted in Selborne and adjoining towns. Similarly, enormous numbers of cereal aphids plagued people in England in 1790 (Curtis, 1845), and in September 1834 an immense cloud of the peach potato aphid swept across the river, covered the quays and streets of Gent, and darkened the sky in both Brugge and Antwerpen (Morren, 1836). Thus, aphids are potentially capable of becoming very abundant over a wide area. Fortunately such plagues are rare. The implication of this is that aphid abundance is normally regulated well below plague levels.

My interest in tree-dwelling aphids started from an observation when experimenting with ladybirds in the field in 1956. Contrary to my expectation the aphid population on sycamore in summer consisted almost entirely of adults. This provoked the question why, and resulted in my studying reproductive aestivation in the sycamore aphid *Drepanosiphum platanoidis* (Schrank), and ultimately the population dynamics of three species of tree-dwelling aphids. That is, tree-dwelling aphids were selected not because they are model organisms for studying population dynamics but because I found them interesting, and having studied under George Varley facilitated my acceptance of the then widely held view, aptly summarized in the statement: 'The world is green'. This implies that herbivores are seldom food limited, appear to be natural enemy limited and therefore not likely to compete for food (Hairston *et al.*, 1960). In addition, because the ladybird *Rodolia cardinalis* had proved such an effective biological control agent, I expected my studies to reveal that ladybirds in particular would be important in the regulation of aphid abundance.

The first aphid I studied was the sycamore aphid. Several students helped me in this task: Robert Russel (1968), Paddy Hamilton (1969) and Martyn Collins (1981) with the natural enemies; Jack Jackson (1970) and David Mercer (1979) with migration; John Shearer (1976) with the other common aphid on sycamore *Pemphigus testudinaceus* (Fernie); Forbes McNaughton (1970) and Peter White (1970) with the effect of the aphid on the growth of sycamore; and Richard Chambers (1979) and Paul Wellings (1980) with quantifying the role of qualitative changes in the aphid in its population regulation. When it became obvious that natural enemies were not important in regulating sycamore aphid abundance I also started to study the lime aphid *Eucallipterus tiliae* (L.). This aphid was included in the study because I saw large numbers of ladybird larvae pupating on the trunks of lime trees in years when lime aphids were abundant. As with the sycamore aphid several students helped me: David Glen (1971) and Steve Wratten (1971) with the natural enemies; Margaret Brown (1975) and Neil Kidd (1975) with quantifying the role of qualititative changes in the aphid; the late Mike Llewellyn (1970) and Peter White (1970) with the effect of the aphid on the growth of lime; and the late Nigel Barlow (1977) with a simulation study of this aphid's population dynamics. Large numbers of ladybird pupae on the foliage of Turkey oak similarly drew my attention to the aphid *Myzocallis boerneri* Stroyan, living on this tree. It was also sampled, along with the continuing census of sycamore and lime aphids, for four years in Glasgow. On my moving to Norwich in 1974 the population census study of the lime and sycamore aphids was discontinued but that of the Turkey oak continued for another 19 years. Aulay McKenzie and Richard Sequeira assisted me with the study of this aphid.

The analysis of the sycamore aphid population census data owes much to Richard Chambers, who introduced me to the concept that the rates of development and growth could be differentially affected by temperature and food quality, and both Richard and Paul Wellings for demonstrating the effect of density-induced qualitative changes in the aphid. That of the lime aphid was very dependent on the simulation study of Nigel Barlow (Barlow & Dixon, 1980), and that of the Turkey oak aphid on the auto-regression analysis of the census data by Richard Sequeira (Sequeira & Dixon, 1997). In terms of the overall synthesis, however, it is a pleasure to acknowledge Pavel Kindlmann's enormous contribution to the understanding of aphid population dynamics and biology. He is a great listener and excellent at expressing ideas in mathematical terms. Seamus Ward, similarly, made an equally important contribution to the analysis and above all the understanding of tree aphid population biology, in particular risky dispersal (Chapter 8) and seasonal sex allocation (Chapter 9). Finally, Vojta Jarosik made a great and valued statistical contribution in exploring the role of density independent processes in the regulation of the Turkey oak aphid.

Starting with the population census of sycamore aphid and then adding the lime aphid, and finally the Turkey oak aphid in my quest for support for the view that natural enemies regulate the numbers of tree-dwelling aphids proved enlightening. The fact that in all three cases, natural enemies did not regulate aphid abundance but density dependent induced changes in the aphids apparently did, convinced me of the generality of the phenomenon. This was necessary because the climate of opinion favoured natural enemies, particularly parasitoids. Although the Natural Environment Research Council (NERC) supposedly favoured long-term population studies, it proved very difficult to obtain funding to do this work. On one occasion a NERC 'visiting group' questioned us about the progress of our research. Although it was not clear why we were so privileged we were left with the feeling that it might be because of the minor role we attributed to natural enemies. The visiting group recommended continued support for the long-term census work providing there was greater emphasis on the effect of aphids on tree growth. In retrospect this directive diverted attention from the main problem – what regulates the abundance of these aphids? Similarly, the time and effort my group subsequently spent studying cereal aphids was also largely wasted. In the early 1970s, Bill Murdoch suggested I apply my ecological understanding of aphids to a more practical problem. A series of outbreaks of cereal aphids resulted in pressure being placed on politicians by the farming community, which resulted in more of the government funding given for agriculture research being used for studies on cereal aphids (Dixon, 1987a). That is, the moral pressure and the availability of funding made it easy for me to work on cereal aphids. Unfortunately, the enormous research effort expended on these studies by several groups lacked a clearly defined applied objective and the availability of funds was determined by political rather than scientific criteria.

As stated by Andrewartha and Birch (1954) population ecology should be based on the study of living organisms in their natural environments. The theoretical framework of population ecology has been more intensively studied than most other aspects of ecology, but is still far from complete. Advances are dependent on insight and imagination, followed by a mathematical formulation of the process under consideration. The resultant mathematical models may reveal other patterns, which were not previously appreciated. It is important that these ideas should be tested experimentally as 'whether or no anything can be known, can be settled not by arguing but by trying' (Bacon, 1620). However, the tendency to present studies in the form of an hypothesis-testing exercise often lacks conviction and rigour. Is it not possible to arrive at an understanding without hypothesis testing – it is reputed that Newton did! It is possibly more important to ask an impertinent question as by so doing one is on the way to a pertinent answer (Bronowski, 1973).

The importance of using long-term population censuses to test ecological theory is widely acknowledged and frequently stressed. However, such studies are tedious to do, difficult to sustain for long periods and initially very unproductive. The methodology of such studies also rarely escapes destructive criticism from population theorists. Therefore, it is not surprising that there are very few long-term studies of insect populations. In addition, convincing others that the accepted dogma might be wrong is far more difficult than to conform. If one assumes all that is necessary is to present the evidence and leave truth to persuade, then one is being very naïve about people's motives. The establishment of the day, which in the case of insect population dynamics favours regulation by natural enemies, is a powerful force with considerable inertia. It is likely in this case that monstrous certainty has impeded rather than aided the search for truth.

Early in 1974 George Varley sent me a copy of his book: *Insect Population Ecology* and in the accompanying letter wrote: 'The aphid population problem, with overlapping generations, is of a different order of complexity from the simple things we now are beginning to understand'. However, if one accepts that the life cycle of an insect can be viewed as beginning and ending with an egg then the supposed complexity largely disappears. On hatching from an egg an aphid clone grows by parthenogenesis and at the end of the vegetative season switches to sexual reproduction and produces eggs. The fact that the 'body' of each aphid clone does not exist in space as a discrete unit but may be scattered widely in space and always consists of many units does not invalidate the life cycle concept. Parthenogenesis enabled aphids to achieve the prodigious rates of increase alluded to above, which have been a major factor in shaping their population dynamics. The objective of this book is to present the case for self-regulation rather than top–down regulation in this group of plant-sucking herbivores. That is, the aphid population problem can be greatly simplified by ignoring the complexity of aphid life cycles. This can be done by focusing on the relationships between the numbers at the beginning and the end of the different phases of increase and decline in a season, and between the end of one season and the beginning of the next. This simple approach was adopted because it resulted in a better system of prediction than more complicated approaches, and was arrived at only after achieving some understanding of the life cycles of aphids.

2

Tree-dwelling aphids

This book mainly deals with the population biology of deciduous tree-dwelling aphids, in particular, the sycamore aphid *Drepanosiphum platanoidis*. Information on the lime aphid *Eucallipterus tiliae* and Turkey oak aphid *Myzocallis boerneri*, which have not been studied in the same detail, is also presented. These aphids all belong to the subfamily Drepanosiphinae, fossils of which are first recorded in the late Cretaceous and make up a large proportion of the early Tertiary aphid fauna. Extant species of this subfamily are mainly host specific and most live on trees belonging to the Fagaceae, Ulmaceae, Aceraceae and Betulaceae. This subfamily of aphids is taxonomically well defined and not as polymorphic as the Aphidinae and Lachninae, which make up most of the present-day aphid fauna. From an ecological point of view this is important as it means they are easily identified, even in the field.

LIFE CYCLE

Most aphids live on only one species of host plant. The sycamore, lime and Turkey oak aphids belong to this group. They spend the winter months as eggs. In spring these hatch and give rise to nymphs that develop into the winged adults of the first generation, known as fundatrices. These adults are parthenogenetic and viviparous, and their offspring develop into other parthenogenetic viviparae. Several parthenogenetic generations occur in succession until the onset of autumn when the nymphs develop into unwinged egg-laying females and winged males. This is the sexual generation, which mate and produce the overwintering eggs (Figure 2.1). Bonnet (1745) was the first to show that aphids may propagate without fertilization for as many as 10 generations. Then, under certain conditions, winged or wingless males appear and copulate with wingless egg-laying females. This

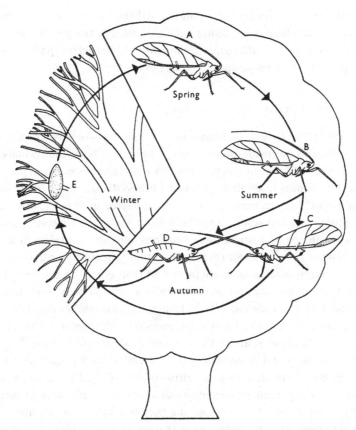

Figure 2.1. Life cycle of the sycamore aphid (A, fundatrix; B, alate
virginopara; C, male; D, ovipara; and E, egg).

cyclical parthenogenesis, in which periods of parthenogenetic reproduction alternate with sexual reproduction is thought to have evolved
in a seasonal climate, possibly that associated with a glacial period in
the Lower Permian. However, live-bearing or viviparity, another characteristic feature of aphids, must have evolved later as the ancestors of
aphids, and the closely related adelgids and phylloxerids are egg-laying
or oviparous (Heie, 1967).

The occurrence within a species of different forms or morphs:
unwinged and winged parthenogenetic viviparae, males and egg-laying
sexual females, is also characteristic of aphids. This polyphenism
reaches its greatest development in host-alternating species of aphids,
which may have as many as eight morphs that differ at least in their
external morphology. Most deciduous tree-dwelling aphids have far

fewer morphs, usually only winged parthenogenetic females, winged males and wingless egg-laying females, although the parthenogenetic females may show differences in external and internal morphology associated with generation-specific strategies (p. 28).

EVOLUTIONARY INDIVIDUAL

As the aphids that hatch from overwintering eggs are parthenogenetic, populations are made up of clones – the 'evolutionary individuals' in Janzen's (1977) terminology. How a clone allocates resources to particular functions is likely to determine its fitness (Dixon, 1977). Individuals in each clone are involved to a varying degree in defence, dispersal, reproduction and aestivation/hibernation (Figure 2.2). However, specialization in one or other of these functions imposes constraints in terms of resource allocation for carrying out other functions. At certain times particular functions are more important than others for the overall fitness of a clone, and this has resulted in a division of labour within a clone, which is reflected in its polyphenism. This implies that altruism is common in aphids. However, the general view is that only the soldier caste is altruistic. That most soldier aphids are sterile and are likely to die defending and safeguarding the survival of their clone mates supports the claim that they act altruistically. Winged aphids similarly have a greatly reduced fecundity and a very low probability of surviving to reproduce, but in dispersing they benefit the overall fitness of their clones. Thus it would appear that winged forms also act altruistically. If this is true, then contrary to Hamilton's (1987) claim altruism is common in aphids.

TELESCOPING OF GENERATIONS

As a parthenogenetic egg does not require fertilization, it can begin to develop as soon as it is ovulated. Under congenial conditions aphids take approximately one week to develop from birth to maturity, whereas other similar sized insects take approximately three weeks. However, if one takes into consideration that an aphid starts developing inside its grandmother (Figure 2.3), then the actual development time is 2.5 times longer than it takes an aphid to develop from birth to maturity, i.e. approximately three weeks. Comparison of the generation times of organisms of a wide range of sizes indicates that the larger and more complex have much longer generation times than smaller organisms (Bonner, 1988). This trend suggests that organisms the size of aphids

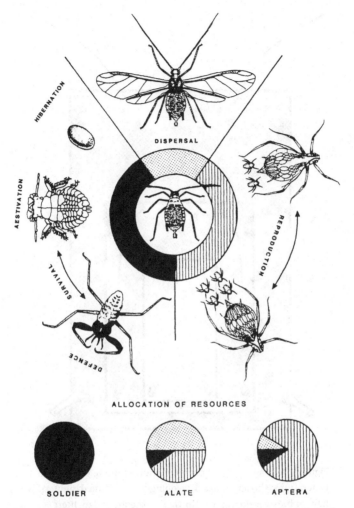

Figure 2.2. Functional aspects of polyphenism and the allocation of resources to dispersal (dotted area), reproduction (hatched area) and survival (filled area) by winged (alate), unwinged (aptera) and aestivating, hibernating and soldier morphs of aphids. (After Dixon, 1985.)

should have generation times in the order of one month and mites in the order of one week (Figure 2.4).

Therefore, there appears to be a minimum 'time' required for development, which is a function of the size/complexity of organisms. Parthenogenesis has enabled aphids to start developing at ovulation and more importantly inside immature or even embryonic mothers. Then, given that there is a constraint on the rate of development, there

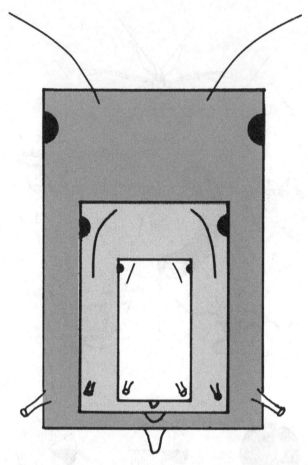

Figure 2.3. Diagrammatic representation of the telescoping of generations; the aphid has developing within it its daughters, which have aphids developing within them – the granddaughters.

are great advantages in telescoping generations (Kindlmann & Dixon, 1989), which has given aphids approximately a threefold reproductive advantage (Dixon, 1990a). In addition, in being viviparous, aphids avoid the heavy mortality experienced in the egg stage in other insects. In this way aphids have been able to achieve rates of population increase normally associated with much smaller organisms such as mites.

Their short generation time also enables aphids to track very closely the seasonal changes in their resources. Individuals in each generation must be able to survive the worst conditions they are likely to

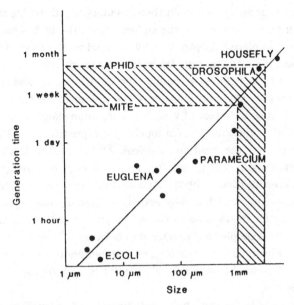

Figure 2.4. The relationship between generation time and size. (After Bonner, 1988.)

experience, which in long-lived individuals is likely to constrain their performance when conditions are favourable. Aphids generally are not so constrained (see p. 28). The seasonal sequence of short-lived generations have generation-specific strategies, by which they anticipate in terms of morphology and physiology the seasonal changes in conditions (p. 28). This close tracking in time, and matching morphology and physiology to seasonal changes in resources, is important in determining the great abundance of many species of aphids relative to their resources.

DEFENCE

Aphids are usually thought of as soft-bodied and defenceless. However, on closer study many show quite complicated defensive behaviour. Depending on the relative size and speed of movement of an approaching natural enemy an aphid will attempt to evade it by kicking, walking away or dropping off the plant. The lime aphid is even capable of jumping backwards, often just as it is about to be caught by a natural enemy. If their attempts at avoidance fail and they are captured aphids can resort to a process called waxing. Predators usually initially seize a hind leg of an aphid, which then attempts to pull its leg free. If this fails it tries to use one of the siphunculi on the fifth segment of its abdomen to place a droplet of waxy material on the head of the predator. The siphunculi in most Drepanosiphinae are relatively long and flexible at their base. By a combination of twisting its abdomen and siphunculi the aphid positions one of the siphunculi, at the tip of which there is now a droplet of waxy exudate, immediately over the top of the head of the predator. On touching the predator's head the droplet spreads over the head and hardens. At this precise moment the aphid attempts to pull its leg free and in a high percentage of cases it is successful (Dixon, 1958). Contained in the waxy exudate is an alarm pheromone, which rapidly evaporates. Aphids close by cease feeding and scatter in response to the alarm pheromone or a combination of the struggle plus alarm pheromone (Dixon & Stewart, 1975). In addition, some aphids are distasteful or even poisonous to predators and others are protected from natural enemies by ants (Dixon, 1958; Rana *et al.*, 2002).

That is, aphids far from being soft-bodied and defenceless are quite effective at avoiding being caught by predators or parasitized. Compared to other aphids many tree-dwelling species are very active and particularly good at avoiding natural enemies. A consequence of this is that the survival of the early stages of their insect predators and the success of their parasitoids is dependent on the early instars of the aphid being very abundant.

FOOD

Aphids scan the surface of a plant accepted as a potential host with the tip of their proboscis by which they detect the contours of veins, their preferred feeding sites (Tjallingii, 1978). They then probe into the plant with their mandibular and maxillary stylets, which together form a hollow needle-like structure (Figure 2.5). The stylet pathway is intercellular, penetrating either through the middle lamella between cells, through secondary wall material, through intercellular air spaces or between the membrane that surrounds the protoplasm of a cell and the cell wall, i.e. intramural/extracellular. However, although the stylets frequently puncture the walls of cells bordering the stylet track they remain undamaged. The highest incidence of punctures is in the cells of the vascular bundles, which indicates that phloem elements may be located by some sampling procedure. Phloem sap is under 15–30 atmospheres of pressure, sufficient to force sap through the extremely fine

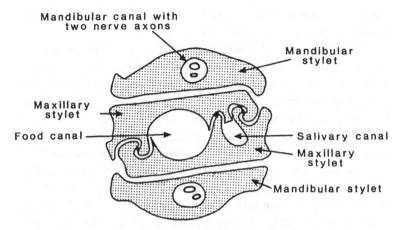

Figure 2.5. Diagram of a transverse section through the stylets of an aphid.

food canal in the stylets and into the aphid's alimentary canal. However, aphids control, by means of valves, the flow of sap into their alimentary canal.

Phloem sap consists mainly of a concentrated solution of simple sugars and a weak solution of amino acids. Although there is little or no need to digest this food as it is already in a soluble form, the nature of the food presents physiological problems. Its osmotic pressure can be much higher than that of the aphids' blood and therefore they are in danger of desiccation, and the low concentration of amino acids would appear to put a severe constraint on the rate of growth they can achieve. Aphids can substantially reduce the osmotic concentration of the ingested fluid, as they excrete honeydew that is similar in osmotic concentration to their blood. They do this mainly by converting mono- and disaccharides into trisaccharides, such as melezitose and oligosaccharides, and thus reduce the number of molecules in solution and the osmotic concentration of their food. It is also possible that they use their so-called 'filter-chamber', a region of the hind gut that encloses the stomach in many aphids, to dilute rather than concentrate the incoming food. Because of the conversion of simple into complex sugars, the liquid in the hind gut is likely to have a lower osmotic pressure than that in the fore gut. This would enable aphids to dilute the incoming fluid by drawing water from their hind gut rather than from their blood. The presence of a capacious stomach in an organism such as the aphid, which feeds continuously, is surprising, but its function may also be to dilute rather than store the incoming food. This is possibly also facilitated by the oesophagous opening into the centre of the stomach rather than at the anterior end.

In contrast to that of sugar, the concentration of nitrogen in the food is relatively very low. As the symbionts of aphids (see below) do not fix atmospheric nitrogen (Smith, 1948), aphids are entirely dependent on their dietary nitrogen (Mittler, 1958). In order to fuel their high rates of growth (pp. 8–10) aphids need to process large quantities of food and use the nitrogen it contains efficiently. Adult sycamore aphids process at least their own weight of phloem sap per day and immature aphids several times their weight (Dixon & Logan, 1973). Compared to Lepidoptera (30–40%; Slansky & Feeny, 1977), aphids are very efficient at assimilating nitrogen from their food, which they achieve with an efficiency of 60% or greater (Mittler, 1958). This is achieved in spite of having to process a relatively large volume of food and apparently in the absence of a mechanism for concentrating the dietary nitrogen prior to assimilation. This is an aspect of aphid physiology that needs further study.

SYMBIONTS

Most insects that live on nutritionally unbalanced diets possess symbiotic micro-organisms (Trager, 1970). In aphids these symbionts are to be found in a structure called a bacteriosome. The primary symbiont in most aphids is the bacterium *Buchneria aphidicola*, which belongs to a subdivision of the γ Proteobacteria that includes *Escherichia coli* and many other intestinal bacteria, including those of modern aphids (Harada & Ishikawa, 1993). Thus it is likely that the ancestor of *B. aphidicola* was a gut microbe. Molecular data and fossil evidence indicate that this symbiont was acquired some 160–280 million years ago (Moran & Baumann, 1994). Many aphids also host secondary symbionts. For example, *Yamatocallis* of the subfamily Drepanosiphinae, a close relative of the sycamore aphid, hosts a specific secondary symbiont, which also belongs to the γ Proteobacteria. Molecular evidence indicates that the ancestors of this aphid acquired this symbiont some 13–26 million years ago (Fukatsu, 2001).

During embryogenesis, symbionts are transmitted transovarially into the bacteriosome of each embryo (Buchner, 1965). At birth the bacteriosome consists of two longitudinal masses of syncytial tissue situated in the body cavity of the thorax and abdomen of an aphid. This structure partly surrounds the alimentary canal and runs from the posterior part of the thorax to the centre of the abdomen and is bounded dorsally by the ovaries. The individual bacteriocytes, which make up the

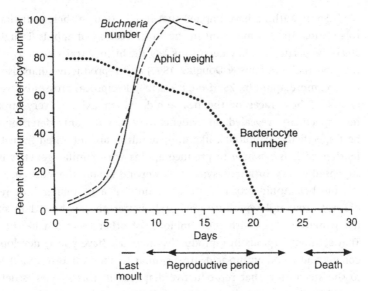

Figure 2.6. Trends in number of symbionts (*Buchneria*) and aphid size, expressed in terms of percentage of the maximum, and number of bacteriocytes relative to the age of an aphid. (After Douglas & Dixon, 1987.)

bacteriosome in adult aphids, are spherical and have a white opaque appearance. Each bacteriocyte contains a peripheral nucleus and its scanty cytoplasm is completely filled with spherically shaped bacteria. During larval life the bacteriocytes increase in size, whereas the bacteria increase in number but remain the same size (Ponsen, 1972). During the latter half of an aphid's nymphal development the number of bacteriocytes starts to decrease but the volume of those that remain increases, which suggests that they and their increasing symbiont population are essential for embryo development. Towards the end of an aphid's reproductive life, bacteriocyte numbers decrease markedly and virtually none remains when an aphid ceases reproduction (Douglas & Dixon, 1987) (Figure 2.6).

This suggests that the symbionts are essential for aphid growth. Physiological studies indicate that they upgrade the amino acids present in an aphid's food and recycle its nitrogenous waste. That is, although their food consists of a weak solution of mainly unessential amino acids, aphids can nevertheless maintain a very high rate of increase by processing large quantities of plant sap and efficiently using the nitrogen they extract. The symbionts undoubtedly play an important role in this efficient use of nitrogen (Buchner, 1965; Douglas, 1989).

Some authors have implied that in addition to their nutritional role symbionts are important in the wider ecology of aphids (Douglas, 1995); in particular they determine key life-history traits such as host race specificity (Adams & Douglas, 1997) and reproductive diapause in the sycamore aphid (p. 28) (Douglas, 2000). Reciprocal crosses between clones of host races of the pea aphid, which exhibits very strong host specificity, revealed no evidence of the symbionts determining host specificity. Symbionts, like mitochondria, are inherited from the mother so it is possible to produce aphids with similar genetics but supposedly very different symbionts depending on the host race of the mother. Aphid crosses with symbionts of a particular host race did not grow and reproduce noticeably better on this host than similar genetic crosses with the symbionts of other races (Knabe, 1999). That sycamore aphids in reproductive diapause have poorly developed gonads and blood amino-acid titres similar to aposymbiotic aphids led to the suggestion that reproductive diapause in this species is determined by the symbionts (Douglas, 2000). Earlier Dixon (1963) recorded that not only were the gonads poorly developed but an endocrine gland, the corpus allatum, in diapausing individuals was smaller and possibly less active than in reproducing individuals. That is, in common with other insects there is evidence that the brain of D. platanoidis can inhibit the activity of the corpus allatum, the secretions of which are known to control reproductive activity in other insects (Johansson, 1958). As the length of reproductive diapause appears to be also dependent on population density (p. 28), it is likely that a higher centre, like the brain, integrates the information so that aphids can respond appropriately. That is, although a programmed response, reproductive diapause varies in intensity and duration, and appears to be an integrated response to nutrition, population density and temperature. If correct then the challenge for physiologists is to determine just how some aphids, such as the sycamore aphid, are able to control the activity of their symbionts.

Further support for aphids having a greater control over their symbionts than vice versa comes from studies on the morphs of aphids, such as males and non-reproductive soldiers, which lack symbionts. As the symbionts are only transmitted through the female line it is tempting to suggest that symbionts selectively infest the embryos that will become reproductive females and avoid those that are destined to become males or sterile soldiers – the symbiont selection hypothesis. Alternatively, the host aphid controls the infection by symbionts – the host selection hypothesis (Fukatsu & Ishikawa, 1992).

The little evidence there is tends to favour the host selection hypothesis. During the blastoderm stage of embryogenesis in aphids a well-defined group of cells, the presumptive bacteriosome, first becomes apparent and just prior to gastrulation the cells of this structure are invaded by symbionts from the mother's bacteriosome (Tóth, 1937). In the case of those species such as *Pemphigus* and *Stomaphis* that have non-feeding (arostrate) males, the male embryos do not develop a presumptive bacteriosome. In the non-reproductive soldiers and males of the aphid *Colophina arma*, and the males of the closely related mealybug *Planococcus lilacinus*, and scale insect *Stictococcus diversiseta*, the symbionts invade the embryo but subsequently degenerate (Buchner, 1955; Fukatsu & Ishikawa, 1992; Tremblay & Ponzi, 1999). That is, embryos appear to control whether or not they are infected by symbionts. The presence of symbionts in males and soldiers that feed and grow (Uichanco, 1924; Ponsen, 1991; Ito, 1994) but do not pass on their symbionts also support the host selection hypothesis. That is, this symbiosis is likely to have costs as well as benefits for aphids and thus it is advantageous for them to control whether they will host and the size of the population of their endosymbionts.

In summary, deciduous tree-dwelling aphids are more host specific and easily identified than most other aphids. Parthenogenesis has enabled aphids to telescope generations, achieve prodigious rates of increase and closely track the seasonal changes in their resources. Although soft-bodied they are highly mobile and very effective at avoiding their natural enemies. Their abundance, however, is very dependent on the way they have overcome the physiological problems posed by the poor quality of their food – phloem sap. They are able to process large quantities per unit time and extract sufficient amino-nitrogen to fuel their very high rates of growth. In this their symbionts play an important role in upgrading the quality and recycling the amino-nitrogen present in their food.

3

Trees as a habitat: relations of
aphids to trees

In this and the following chapter the niche occupied by the sycamore aphid will be described. Elton (1927) was the first to adopt a dynamic approach to this concept. He argued that the external factors acting on an animal must be included when characterizing its niche, which he defined as: an animal's place in the biotic environment and *its relations to food and enemies*. This chapter gives an account of the relations of the sycamore aphid to sycamore, which serves as its home and source of food. In the next chapter the relations of the sycamore aphid to its natural enemies are presented.

DECIDUOUS TREES

Deciduous summer forests tend to dominate in those parts of Europe where summer temperatures are high and air is moist. Within this forest habitat there can be one or several dominant tree species, each of which is host to one or more species of aphid, most of which are highly host specific. That is, trees constitute the biotic environment of this group of aphids. Their niche is that of a relatively small herbivore feeding on the leaves of trees, which make up the forest canopy 'one of the least explored zones on land' (Elton, 1966).

The dominant species of tree in this habitat is often oak (*Quercus* spp.). Unlike oak, lime (*Tilia* spp.) and sycamore (*Acer pseudoplatanus*) trees rarely dominate over large areas. Although now widely distributed because it is a good colonizer, sycamore previously is likely to have been confined to mountainous areas, growing alongside streams and small rivers. Lime is a more frequent component of oakwoods, especially in limestone areas.

Because of their height, trees are difficult to sample and during a long-term study are likely to increase in size quite considerably.

Sycamore

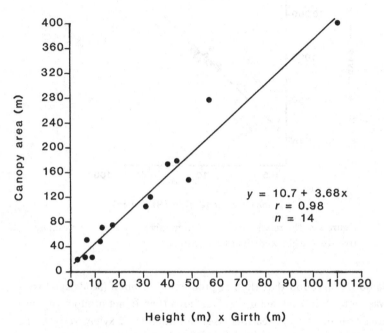

$$y = 10.7 + 3.68x$$
$$r = 0.98$$
$$n = 14$$

Height (m) x Girth (m)

Figure 3.1. The relationship between canopy area and sycamore tree size (height × girth at base of tree).

However, their geometry can be defined and used to estimate aphid abundance and the contribution of the aphids to energy and nutrient fluxes in forests. For example, for sycamore it is possible to obtain an estimate of the canopy area (Figure 3.1) and the number of leaves on a tree (Figure 3.2) by measuring the circumference of its trunk at ground level and its height, which give an indication of a tree's size. In addition, if one assumes that the canopy of sycamore is conical in shape, then using the above estimates, it is possible to determine the number of leaves in the lower, middle and upper third of the canopy, which is 46%, 38%, and 16% of the total. For energy-flow studies it is also useful to have an estimate of the area of the ground covered by a vertical projection of the canopy of a tree. From this and the previous and other relationships it is possible to estimate the biomass of leaves, biomass of aphids, etc., per unit area of ground (pp. 153–4).

The total mass of leaves produced may be more important in determining net productivity than photosynthesis per unit leaf area.

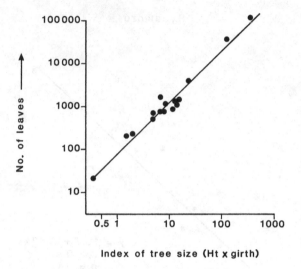

Figure 3.2. The relationship between number of leaves and sycamore tree size (height × girth at base of tree).

Satoo (1970) found that the net production of birch (*Betula* spp.) trees was more closely related to total leaf area than to net productivity/unit area of leaf. As leaf area increases the number of xylem vessels, and thus the cross-sectional area of xylem (wood) supplying the leaves with water, is likely to increase. In sycamore there is a strong correlation between the area of the current year's xylem and leaf area (Dixon, 1971a). Therefore, knowing the dimensions of a tree, in particular the leaf number/area, it may be possible to estimate the net productivity in terms of the volume of wood added each year.

SYCAMORE

Before discussing whether the abundance of an organism is resource or natural enemy limited it is necessary to define its habitat and niche. This will be done for the sycamore aphid because it is the best known of the species. The numbers of this aphid on eight trees, on three sites in Scotland, were monitored weekly. Two of the trees were in the grounds of the University of Glasgow, the city site; two at Bearsden, a suburb of Glasgow, the suburban site; four in mixed deciduous woodland on the Rossdhu Estate, 34 km northwest of Glasgow, the woodland site. Although the other tree-dwelling aphids were not studied in the same detail they are likely to occupy very similar niches.

From bud burst to leaf fall each year tree canopies appear to change very little. However, there are changes that have shaped the life-history strategies of the aphids that live there. At bud burst in spring, leaf growth is sustained by amino-nitrogen imported into the leaves from the trunk. In autumn the reverse process occurs when some of this amino-nitrogen is salvaged from the leaves prior to leaf fall. This flow of amino-nitrogen in and out of the leaves determines the quality of food available to aphids at any particular time. In addition, weather in summer tends to be warmer and the intensity of solar radiation greater than in spring and autumn. Together these changes determine the quality of this habitat for aphids. The marked seasonality in the trend in habitat quality – high in spring and autumn and low in summer – is similar each year. At any one time, however, not all the leaves are equally suitable for aphids. Wind and exposure to solar radiation make the microenvironment on some leaves unsuitable for aphids. That is, relative to their host, aphids are small and as a consequence they live in a very coarse-grained or heterogenous environment (Levin, 1968). The role of this seasonality and environmental heterogeneity in shaping the sycamore aphid's life cycle and way of life is discussed below.

SEASONALITY

Although many of the species of aphids that live on shrubs lay their eggs in crevices close to the dormant buds (e.g. Behrendt, 1963; Way & Banks, 1964) sycamore aphids lay very few, if any, eggs close to the terminal buds. At distances of over 50 cm from the terminal buds the number of eggs per unit area of bark increases with distance (Figure 3.3a) (Dixon, 1976a). That is, most eggs are laid in crevices in the thicker older bark on the main trunk at considerable distances from the buds, especially those in the upper canopies of mature trees. A possible advantage of this is that it affords them protection from birds.

As the eggs are laid well removed from the buds, the regions of high hormonal activity at bud burst, it is unlikely that the embryos within the eggs synchronize their development with that of their host by monitoring hormonal changes in the buds. Both egg hatch and bud burst are associated with the accumulation of day degrees in spring (Figure 3.3b). That is, both appear to use the accumulation of thermal units as a measure of time. There is a very close association between the average time to bud burst and egg hatch for buds and eggs kept at a range of temperatures. Extrapolation of the relationships between developmental rate (1/D) and temperature indicate that the lower

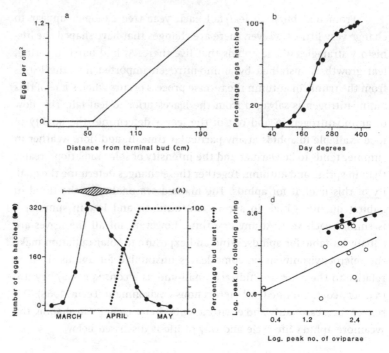

Figure 3.3. Relationships between: a, sycamore aphid eggs per unit area of bark and distance from terminal bud; b, percentage of eggs hatched and day degrees; c, number of eggs hatched and percentage bud burst, and time in spring; and d, the peak number of aphids on leaves at bud burst relative to the peak number of egg-laying females (oviparae) the previous year present on trees that burst their buds early (•) and late (o), respectively. (A), Distribution in time of bud burst of trees in the vicinity of the tree on which egg hatch was studied.

developmental threshold for buds is 1 °C, and for eggs 5 °C. This, and the relationship depicted in Figure 3.4, indicates that although aphid and host possibly use the same mechanism for monitoring the passage of time, the relationships with temperature differ. There are very big differences in the time of bud burst between trees, and trees that tend to burst buds early in one year tend to do so every year and vice versa. If it is advantageous for aphids to synchronize their development with that of their host it is likely they would evolve a mechanism that synchronizes egg hatch with average bud burst of all the trees in the area. That sycamore aphids have evolved such a mechanism is indicated by the fact that average egg hatch and bud burst times are extremely well synchronized (Figure 3.3c).

What is the advantage of the sycamore aphid synchronizing egg hatch with average bud burst, and what prevents it from becoming

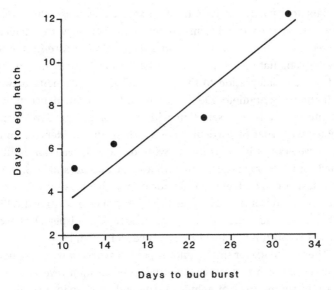

Figure 3.4. The relationship between the average number of days to egg hatch and to bud burst of eggs and saplings kept at different temperatures.

BUD BURST

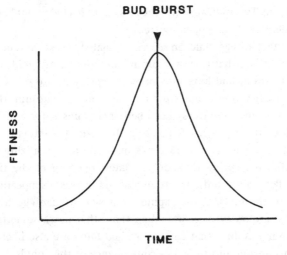

Figure 3.5. Diagrammatic representation of the fitness of aphids hatching at different times relative to bud burst (▼).

even more effective in tracking the development of its host? Close synchronization results in the development of large, quickly maturing and highly fecund adults, with those hatching earlier or later being less fit (Dixon, 1976a; Figure 3.5; p. 24). A high incidence of inter-tree movement (Dixon, 1969; Chapter 8) reduces the degree to which an aphid can adapt to a particular tree. In other species, such as *Kaltenbachiella japonica*, where the winged forms are relatively sedentary, the incidence of inbreeding is high, which leads to a very rapid increase in homozygosity. Sub-populations on individual trees become genetically differentiated through adaptation to the phenological heterogeneity between trees (Komatsu & Akimoto, 1995). Thus, providing the incidence of inter-plant movement is low, selection will synchronize the development of aphids with that of particular hosts. Where the incidence of inter-plant movement is high, as in the sycamore aphid, selection is likely to synchronize average egg hatch with average bud burst (Fig. 3.3c(A)) – i.e. the scale of the adaptation, fine or coarse, depends on the degree of inter-plant movement. However, why some species of tree-dwelling aphids show a high and others a low incidence of dispersal between host trees is puzzling and will be returned to in Chapter 8.

The advantage for the sycamore aphid of synchronizing egg hatch with bud burst can be further quantified. Census data reveals that an increase in the number of aphids on the buds results in a significant and proportional increase in the number of aphids colonizing the leaves at bud burst. However, the longer the period between egg hatch and bud burst the fewer aphids that survive to colonize the leaves (Dixon, 1976a; Figure 3.6). This indicates that the loss of aphids before bud burst can be considerable on late opening trees.

The number of eggs laid on a tree is related to the number of egg-laying females on that tree in autumn. The number of aphids that colonize the leaves at bud burst the following spring is also related to the number of egg-laying females present the previous autumn. However, the relationships for trees that open their buds early and late differ. A peak of 100 egg-laying females on the earliest opening of the eight trees studied gave rise to 13 times more aphids colonizing the leaves the following spring than on the latest opening of the trees (Figure 3.3d). Both birds and rain are implicated in this disappearance of aphids from the buds. Clearly aphids that synchronize egg hatch with bud burst are on average fitter than those that hatch some time prior to bud burst. Aphids that hatch after bud burst are also likely to be less fit. This decline in fitness is a consequence of the aphids developing on older and less nutritious leaves. Overall, therefore, sycamore aphids that synchronize their egg hatch with bud burst are likely to be fitter than those that hatch either prior to bud burst or when the leaves have unfurled (Figures 3.5 & 3.6). That is, in the case of the sycamore aphid there is likely to be a strong selection pressure favouring those aphids that synchronize egg hatch with bud burst. Similarly, the time of leaf fall, which also varies in the same way between trees, will have

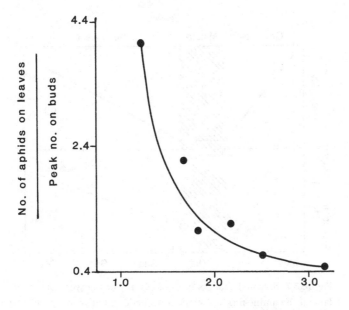

Figure 3.6. The relationship between the proportion of the aphids on the buds that colonize the leaves at bud burst and the period that elapsed between peak egg hatch and bud burst.

determined the time of the switch from asexual to sexual reproduction and the production of overwintering eggs. In the case of the sycamore aphid, where there is a high incidence of between-tree movement, these responses are likely to be to the average times of bud burst and leaf fall rather than the values for particular trees.

The reproductive rate of the sycamore aphid changes markedly during the season, and is strongly correlated with the level of amino-nitrogen in the leaves of sycamore (Figure 3.7). Reproduction is at a high rate when the leaves are actively growing or senescing, and at a low rate, or ceases altogether, when the leaves mature. The increase in the reproductive rate at the beginning of a year occurs because the adults at first are mostly immature and then come into reproduction and finally reach their maximum reproductive rate. In autumn the decline results from the drying out of the leaves, which then no longer proffer a satisfactory food source.

Associated with the changes in reproductive rate there are also changes in the size of the aphid (Dixon, 1966). In spring and autumn the adults are two to four times heavier than they are in summer.

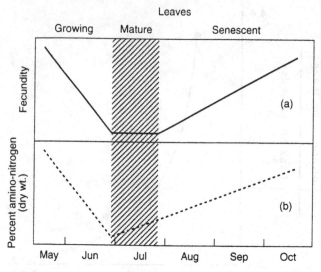

Figure 3.7. Seasonal changes in the fecundity of the sycamore aphid (a) and the amino-nitrogen content of the leaves of sycamore (b). (After Dixon, 1970b).

Under field conditions it is difficult to separate the role of nutrition and interaction between aphids in determining body size. Rearing aphids in isolation from the first instar to maturity on leaves of plants that are either growing, mature or senescing results in adults of different sizes comparable to those observed in the field. Aphids reared in crowds on leaves of the same plant are always smaller than those reared in isolation. Thus the effect of crowding during the development of an aphid, as well as the nutritive quality of the host plant, influences the size of adult aphids (p. 82).

The reproductive performance of an adult varies according to its size. In spring and autumn there is a delay of several days after the adult moult before the onset of reproduction. The duration of this delay is correlated with the size of an aphid. Large aphids come into reproduction sooner than small aphids. The large aphids also give birth to more and larger offspring than small aphids. Larger aphids are also less likely to die before reproducing (p. 88).

Associated with the seasonal changes in reproduction and size are changes in activity. In the case of the adults they are relatively more active early and late in a year than they are in summer (Figure 3.8). That is, when reproducing they frequently move from leaf to leaf possibly searching for more favourable sites for reproduction. In contrast

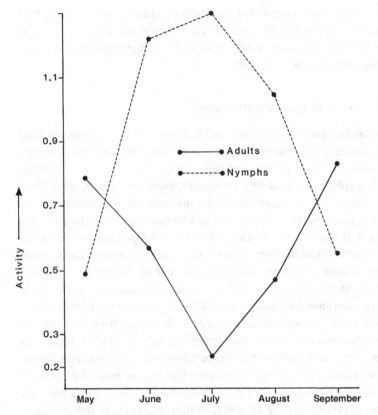

Figure 3.8. The seasonal trend in the average monthly activity of nymphs (1st–3rd instars) and adult sycamore aphids on two trees at one site over a period of three years. Activity was measured as the average number moving on to certain leaves each day divided by the average number per leaf per tree over the same period.

the young instars (Figure 3.8) are most active in the middle of a year when the leaves have the lowest nutritional value. This possibly indicates that for aphids that are still growing and are on mature leaves it is more advantageous to move and seek a more favourable leaf than stay on the same leaf. Certainly any leaves that senesce prematurely in summer are quickly located by immature aphids. Therefore, it is unlikely that branches of a tree that break their buds later than the other branches (Leather, 1996) will remain uninfested for long and so escape in time as suggested by Leather (2000). They are likely to be quickly located and colonized by young and above all adult aphids (see Figure 3.8).

Thus the seasonal changes in the nutritional quality of the host plant have a marked effect on the reproductive rate and activity of its aphids. Good nutrition results in active, large, high quality aphids with a high reproductive potential.

Generation-specific strategies

The second generation adults, which develop and are present in summer, enter a reproductive diapause and have a low fecundity thought to be a direct consequence of the poor quality of the food available to the aphids when feeding on mature leaves (Mordvilko, 1908). However, when reared under the conditions normally experienced by the first generation, the large second generation individuals that develop have well developed fat bodies and poorly developed gonads as if about to enter aestivation. That is, they anticipate the onset of harsh conditions associated with the cessation of growth of the leaves of sycamore (Dixon, 1975b). The duration of aestivation is density dependent, being longest in those years when the aphid is abundant in summer (Wellings et al., 1985). Second generation aphids also differ from those of the first generation in having fewer ovarioles (Wellings et al., 1980), longer appendages and a greater number of rhinaria on their antennae (Dixon, 1974), a longer gut (Dixon, 1975b) and a higher wing beat frequency, and fly for longer (Mercer, 1979). All features that are likely to improve their chances of surviving the adverse conditions prevailing in summer. Interestingly, the closely related Drepanosiphum acerinum, which also feeds on the leaves of sycamore, continues reproducing all through summer. This species hatches after the buds of sycamore have burst, usually when the leaves are fully expanded, produces sexuals and overwintering eggs well before leaf fall and, by preferring to live on sycamore saplings growing in shaded conditions, avoids the direct heating effect of the sun (Dixon, 1998). This might indicate that the two species have different lower developmental thresholds. For the sycamore aphid, which is present when the temperature is quite low in early spring and late autumn, it could be advantageous to have a low developmental threshold. In the case of D. acerinum, which is only active from early summer to early autumn, a higher lower developmental threshold would appear to be more appropriate. If this is correct, and it has some empirical support (Wellings, 1981), then reproductive diapause in the sycamore aphid, which is such a marked feature of its life cycle, could be a consequence of a physiological constraint. The lower developmental threshold is advantageous for exploiting the cooler conditions prevailing in early spring and late autumn but a constraint in the warmer conditions prevailing in summer. This is because the rate of development relative to the growth rate that can be sustained at that time of year would possibly result in the production of very small non-viable adults.

In autumn, mainly in response to short days and lower temperatures, sycamore aphids switch to producing sexual forms, which produce the overwintering eggs. In this way they anticipate leaf fall by switching to a resting form prior to the onset of adversity. In addition to the direct response to environmental cues the aphid has an internal clock, which governs the intensity of the response to the environmental cues. Later generations give birth to proportionally more sexual offspring in response to a particular set of sexual-inducing environmental conditions than the earlier generations (Dixon, 1971d, 1972a). This is discussed in more detail in Chapter 9.

SPATIAL HETEROGENEITY

Vertical distribution

Because of the height of trees it is difficult to sample the leaves in the upper canopy. Therefore, there is a tendency for researchers to sample only leaves in the lower canopy and assume this is representative of the whole tree. This assumption is rarely if ever tested. Of the eight sycamore trees sampled for aphids, one tree was sampled weekly at three heights – bottom, middle and top – for eight years. This revealed that at the beginning of each year the aphid is equally distributed vertically (Figure 3.9). As the leaves mature and the second generation adults enter reproductive diapause the aphids in the upper canopy move down and aggregate on the leaves in the lower canopy. During periods of overcast weather conditions in summer there is some recolonization of the middle canopy, which is quickly vacated when sunny conditions return. Early in August, just as the aphids start to reproduce, they also start to move back up the canopy. Around the middle of September the numbers of aphids per leaf was roughly equal at the bottom and top of the canopy. After this, and for the rest of the season, the numbers of aphids per leaf was greatest in the lower canopy.

In summer, the nutritional quality of the leaves in the upper canopy is similar to that in the lower canopy. However, these leaves are more exposed to solar radiation and, as a consequence, the average temperatures close to the underside of the leaves high in the canopy

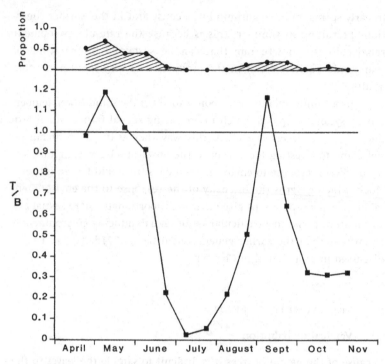

Figure 3.9. The seasonal trend in the ratio of the numbers of aphids per leaf at the top of a tree (T) to those per leaf at the bottom of a sycamore tree (B), sampled weekly over a period of eight years. The upper figure is the seasonal trend in the occupancy of the leaves in the upper canopy expressed as a proportion of that in the lower canopy.

are generally higher than in the lower canopy (Figure 3.10). In summer at midday the difference in temperature can be as much as 10 °C (Figure 3.11). This, combined with the poor nutrition generally available throughout the canopy in the summer, makes the upper canopy an unsuitable place for aphids.

   That the sycamore aphid attempts to avoid heat stress when it is in reproductive diapause in summer is also indicated by their tendency to lack melanic pigmentation in summer. This contrasts markedly with the tendency for them to have a dark-coloured head, thorax, legs and up to eight black bands on the abdomen in spring and autumn (Figure 3.12). Rearing them in the laboratory at different temperatures indicates that the production of melanic pigmentation is triggered by low temperature. The colder the rearing temperature the darker they are as adults. This is advantageous as it enables them to absorb solar

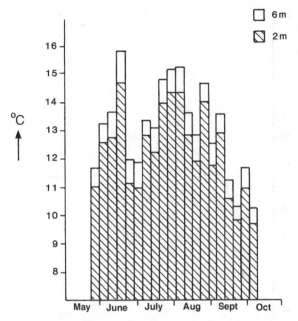

Figure 3.10. Average weekly leaf temperatures recorded at 2 m and 6 m in the canopy of a sycamore tree.

radiation and maintain body temperatures several degrees above ambient when it is cold (Dixon, 1972b). However, in summer when it is warm they do not develop melanic pigment and even in the lower canopy avoid leaves or parts of leaves directly exposed to the rays of the sun (Dixon & McKay, 1970).

The dramatic seasonal changes in the vertical distribution of aphids within the canopy of trees are achieved by adults flying between leaves. The flight activity of the adults was monitored by two traps: a suction trap opening 1.4 m above the ground and positioned 7 m south of a sycamore tree that had a vertical cylindrical sticky trap positioned close to the edge of its canopy 3 m above the ground (Dixon, 1969). The ratio of the numbers of aphids caught by these two traps shows a marked seasonal trend. In spring when the aphids are uniformly distributed throughout the canopy (Figure 3.9) adult aphids predominantly fly between trees and are mainly caught by the suction trap. In summer, the aphids fly down and colonize leaves in the lower canopy and are seen frequently hovering amongst the leaves in the lower canopy. This results in relatively more aphids being caught by the vertical sticky trap. In autumn, the aphid recolonizes the middle and upper canopy,

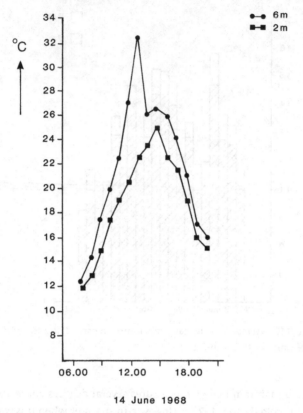

Figure 3.11. The daily trend in the leaf temperatures recorded on
14 June, 1968 at 2 m and 6 m in the canopy of a sycamore tree.

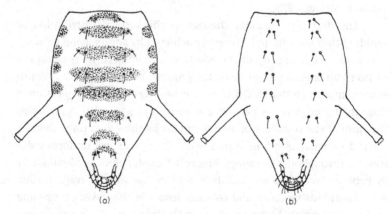

Figure 3.12. The pattern of melanic pigmentation on the dorsum of the
abdomen of sycamore aphids maturing in spring or autumn (a) and in
summer (b).

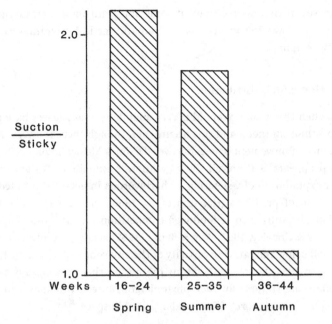

Figure 3.13. The number of adult sycamore aphids caught by a suction trap positioned close to a sycamore tree divided by the number caught by a sticky trap positioned close to the canopy of the same tree in spring, summer and autumn over a period of two years. (The difference between the summer and autumn ratios is highly significant $P < 0.001$; after Dixon, 1969.)

which results in relatively even more aphids being caught by the sticky trap (Figure 3.13). In this way the aphid is able to track the vertical changes in habitat quality, and locate and occupy the most suitable leaves.

The lime aphid also shows a preference for the lower canopy. In May and June the numbers per leaf in the upper and lower canopies are similar, but from then on there is a decline in the occupancy of the leaves in the upper, relative to the lower, canopy. In September there are three times more aphids per leaf in the lower than in the upper canopy (Llewellyn, 1970). This has also been reported for the lime aphid by Dahlsten et al. (1999), and for birch aphids (Hajek & Dahlsten, 1988), who attribute this to the aphids preferring the cooler and shadier parts of the canopies of trees (Dahlsten et al., 1999). The vertical distribution of the oak aphid Tuberculoides annulatus, however, appears to remain homogeneous throughout the year (Lorriman, 1980). That is, although some, but not all, tree-dwelling aphids show a similar preference for the leaves in the lower canopy at some stage during a year, no other species is known to show the same dramatic vertical migration as the sycamore aphid.

Horizontal distribution

Even when the sycamore aphid is very abundant, many leaves have few or no aphids on them, and this occurs even though there is a very high incidence of movement between leaves (Dixon & McKay, 1970; pp. 26–8). Thus it appears as if not all the vegetation is exploited. This apparent under-exploitation of vegetation by herbivorous insects is attributed to the action of predators (Hairston *et al.*, 1960), or aggregation of the herbivores resulting in self-induced competition for food and/or space (Kennedy & Crawley, 1967; Way & Banks, 1967) keeping herbivore numbers well below the carrying capacity of the environment. Other authors (Huffaker, 1957; Voûte, 1957) favour the notion that only certain parts of a plant are suitable for each phytophagous insect, which as a consequence are often short of food and/or suitable space.

Young leaves

Sycamore leaves grow very rapidly after bud burst in spring and all leaf growth is complete by the second week of June. During this period the aphids aggregate on the very youngest leaves at the tips of the branches. Observations tend to indicate that the most heavily infested leaves are on twigs where the internodes are short, resulting in the terminal leaves either nearly or actually touching the sub-terminal pair of leaves. Aphids aggregate on the undersurface of the terminal leaves where they overlap the sub-terminal pair, pocketed in folds of the terminal leaf or in its concavity, and protected by the leaf below. That the aphids prefer leaves with a particular microenvironment can be shown by removing the leaf immediately below one of the heavily infested terminal leaves. Within 5 minutes of removing the lower leaves most of the aphids begin to move about on the terminal leaves and quickly leave them (Figure 3.14); whereas on the adjacent terminal leaves the aphids remain undisturbed. Although it is not possible to discount the effect of nutritional changes resulting from the removal of the sub-terminal leaves, it is highly improbable that these could occur so quickly. It is more likely that the closeness of the two leaves supply a microenvironment favourable to aphids. As this occurs in spring when temperatures are relatively low the enclosed microenvironment might be warmer

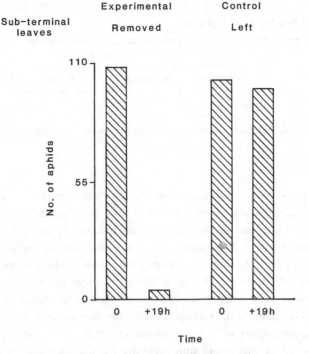

Figure 3.14. The effect of removing or leaving in place the sub-terminal leaf on the number of aphids remaining on the terminal leaf 19 hours later.

and/or offers some protection from predation by birds. Thus, it seems likely that it is advantageous for aphids to select not just the youngest growth but that which has a particular microenvironment associated with it.

Mature leaves

Sampling the same leaves daily for long periods of time indicates that some are consistently infested with aphids, others with few or no aphids and others are more variable, switching back and forth between being lightly and heavily infested with aphids for short periods of time. The ranking of leaves in terms of the average level of infestation is weakly correlated with their size, not associated with nutritional quality, but clearly associated with the position of a leaf relative to others in the canopy. The less suitable leaves more frequently have other leaves immediately below them than do the suitable leaves (Table 3.1).

Table 3.1. *Position of suitable and less suitable leaves relative to other leaves immediately below them in the canopy* $(\chi^2 = 7.3, P < 0.01)$

	Overlap			
Leaf	Complete	Large	Slight	None
Suitable	4	2	4	16
Less suitable	15	3	1	6

Wind

Aphids use their tarsal claws, and in the case of the Aphidinae pulvilli, pliable cuticular sacs, everted from the tibia tarsal articulation (Lees & Hardie, 1988), and in Drepanosiphinae, sponge like empodial pads positioned between their claws, to adhere to surfaces (Kennedy, 1986). Even on an inverted smooth surface the force with which these pads and sacs adhere to the surface is about 20 times greater than the gravitational force tending to detach each foot of the aphid. Aphids secrete a liquid on to the surface of their adhesive pads and sacs, which are in close contact with the substrate, and it is likely that the surface tension of this liquid film accounts for the adhesion. In addition to being able to adhere to smooth surfaces, aphids often have to avoid natural enemies, which involves rapidly detaching themselves from the substrate. An advantage of using a surface-tension-based adhesive mechanism is that by tilting the attached surface, the force required to remove (peel) the surface from the substrate is considerably reduced (Dixon *et al.*, 1990). That is, although an aphid is able to move quickly, nevertheless, it is very difficult to blow it off a smooth surface.

Mature leaves are born on long flexible petioles and are easily moved by wind. The mass of a leaf relative to that of an aphid is very great – of the order of 2 g:1 mg, a ratio of 2000:1. Thus, an aphid struck by a leaf touching the underside of the leaf on which it is feeding is likely to be dislodged. Bending leaves, so that the undersurface becomes concave rather than flat, makes it less likely that aphids will be brushed off by other leaves, and placing leaves immediately below leaves in the canopy of sycamore trees in the field dramatically affects their suitability for aphids. The average ranking of bent leaves improves significantly relative to the control leaves. Similarly, positioning a leaf immediately below a leaf results in a rapid and dramatic decrease in the number of aphids on that leaf, and its subsequent removal in a gradual increase

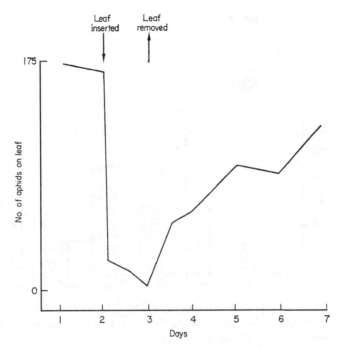

Figure 3.15. The effect on the number of aphids on a leaf of placing
another leaf immediately below it.

in the numbers of aphids on the leaf (Figure 3.15). In the still conditions of a greenhouse the suitability of a leaf is not affected by placing
another leaf immediately below it. Thus the suitability of a leaf in the
field is related to the presence of a leaf immediately beneath it, which
in windy conditions brushes against its underside.

Sticky traps placed beneath the canopy of sycamore trees intercept immature aphids falling from the canopy (Figure 3.16). The number falling off is correlated with the numbers present on the leaves
and the incidence of movement between leaves. Statistical analyses in
which these factors are kept constant reveal that there is a significant
correlation between the number of aphids falling from a tree and wind
speed, regardless of nymphal instar, but is particularly marked for the
larger fourth instar individuals. This may not be because they are being
blown off the leaves but because they are more exposed to the brushing
action of leaves. The larger leaf veins protrude from the undersurface
of a leaf and thus for small aphids, which have a tendency to settle
alongside veins, this affords some shelter from the brushing action of
another leaf. The larger the aphids the less likely they are to obtain

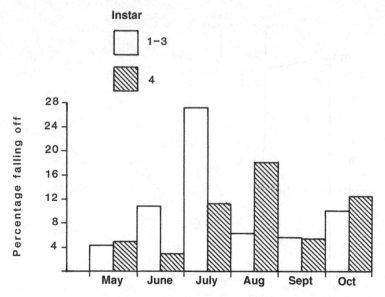

Figure 3.16. The average percentage of first to third instars and fourth instar aphids dislodged or falling off the leaves in the lower canopy of two trees over a period of two years.

protection from brushing by settling alongside veins, as only the very largest veins protrude far enough to afford protection for them.

As the population increases in number, fewer and fewer leaves of a sycamore tree remain uninfested with aphids. Thus, with increasing population density it is likely that an increasing proportion of the aphids will find themselves on less suitable leaves or parts of leaves (p. 34). If this is correct, then a wind of the same strength will dislodge proportionally more aphids from trees where aphid population density is high than where it is low. This is supported by the following observation. The day before Hurricane Faith struck Glasgow on 6 September 1966, three sycamore trees were sampled. The hurricane lasted for four days after which the trees were sampled again. Proportionally more aphids disappeared from the heavily infested trees (Figure 3.17). That is, high winds can dramatically perturb population increase, especially when aphids are abundant. In high winds most of the aphids on a leaf aggregate at the junction of the leaf lamina with the petiole (Figure 3.18). At this point the leaf veins are large and protrude well below the undersurface of the leaf thus affording considerable protection to the aphids. In very high winds many of the aphids move off the leaves on to the branches, which become noticeably green

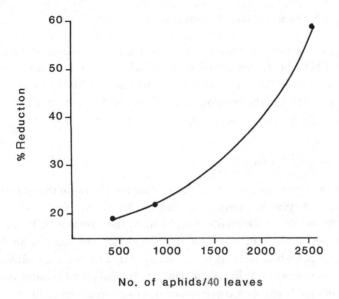

Figure 3.17. The percentage reduction in the numbers of aphids on three sycamore trees relative to the numbers present on the trees before the start of Hurricane Faith (Glasgow 3–6 September 1966).

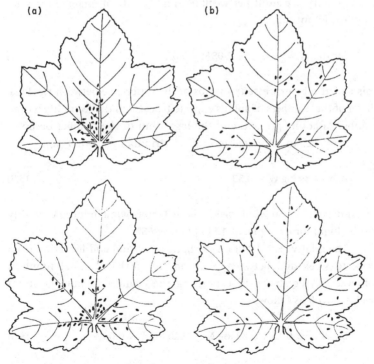

Figure 3.18. The distribution of aphids on two leaves (a) during a high wind (Hurricane Faith) and (b) after the high wind abated.

because the masses of aphids aggregating there. Even in very high winds, both in the laboratory and the field, aphids were never seen blown off a leaf. Therefore, the shelter afforded by the large leaf veins and moving on to the branches is not from the direct effect of wind so much as from leaves brushing against one another.

PERTURBATIONS

With few exceptions, the size of the autumn population of the sycamore aphid each year is inversely related to the spring population size (pp. 70, 86). The peak numbers achieved in the autumns of 1960 and 1972 were greatly in excess of that predicted. The weather in the autumn of both these years was unusual in that the wind speeds were well below average. As indicated above, high wind dislodges large numbers of aphids and as a consequence is an important mortality factor. Therefore, in still autumns this aphid is likely to become extremely abundant.

The effect of average wind speed in autumn (W), and the average fecundity per adult per week (F), on the rate of increase in aphid numbers in autumn is:

$$\log(N_{t+1}/N_t) = 0.11 + 0.05F - 0.03W \qquad (3.1)$$

This can also be expressed in terms of the difference between the logarithm of the numbers of aphids expected, based on the fecundity (F) and number of adults present, and the logarithm of the number observed each week, i.e. an estimated mortality attributable to wind (k_w).

$$k_w = 0.89 \log W - 0.53 \qquad (3.2)$$

This indicates that at wind speeds above 3 knots wind-induced mortality can be high; for example, at 12 knots it is 65%.

The relative contribution of changes in wind and fecundity from week to week and between years on the peak autumnal numbers achieved can be shown by expressing the changes in numbers from week to week as follows:

$$\log N_{t+1} = 0.15 + 0.83 \log N_t + 0.25 \log R - 0.03W \qquad (3.3)$$

where R is F multiplied by the average number of adults present. This reveals that both wind speed and recruitment markedly affects

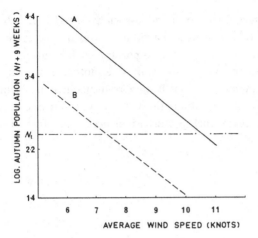

Figure 3.19. Relationships between average wind speed in autumn and peak populations achieved in autumn predicted by equation (3.3) when the autumnal reproductive rates are good (A, 1968) and poor (B, 1973). Populations at the beginning of each autumn (N_t) are 300 aphids. The numbers plotted are those present nine weeks later.

autumnal peak numbers (Figure 3.19). That is, in addition to the operation of a delayed density dependent effect on fecundity (p. 86), wind speed is acting as a perturbing factor. A drop in average wind speed from 9 to 6 knots can result in a 15-fold increase in the peak number achieved (Dixon, 1979).

In summary, tree-dwelling aphids live in a very seasonal and coarse-grained geometrical environment, which is nutritionally very favourable in spring and autumn but less so in summer. That each generation of aphids lives for only a short period and shows generation specific strategies, enables them to track closely the seasonal development of their host. However, adapting to the cool conditions in spring and autumn may have constrained the sycamore aphid's ability to continue developing and reproducing in summer, which possibly accounts for the summer reproductive diapause and the vacating of the upper canopy in summer so characteristic of this species. Synchronization of egg hatch with bud burst and sexual reproduction with leaf fall is clearly advantageous but as there is a high incidence of between tree movement, these responses are to average time of bud burst and leaf fall rather than these values for particular trees. The fact that many leaves consistently have few or no aphids on them, even when the aphid is abundant, is mainly due to differences in the microenvironment associated with leaves rather than nutrition. In this the brushing action of leaves and direct exposure to solar radiation are the most important. High winds agitate the leaves and cause them to brush against one another, converting a favourable environment into a hostile one, resulting in aphids being dislodged from leaves and a dramatic switch from positive to negative population increase. Such perturbations occur most frequently in autumn.

4

Trees as a habitat: relations of aphids to their natural enemies

An abundance of prey generally attracts large numbers of a diverse array of natural enemies. This is very true of the sycamore aphid, which is attacked by a large number of parasitoids and insect predators (see Table 6.4) that differ in taxonomy and tactics. As they all feed on the sycamore aphid, then following Root (1967), they constitute a guild. In jointly exploiting a patch of prey the members of a guild may affect one another's foraging success. This interaction, however, differs from competition in that one participant, the predator, accrues immediate energetic gain as well as reducing potential competition. Because of the way this guild is structured the most important interactions are likely to occur between the immature stages of the natural enemies and the parasitoids are more likely to be adversely affected than vice versa. Below is an account of the most frequently encountered natural enemies of the sycamore aphid.

PARASITOIDS

The larval stages of parasitoids develop either internally or externally upon a single host, the latter eventually dying as a result of the attack. The adults are, with few exceptions, free-living and their food source is usually distinct from that of the larvae.

Primary parasitoids

The sycamore aphid has four specific primary hymenopterous parasitoids: the chalcid *Aphelinus thomsoni* Graham and the braconids *Monoctonus pseudoplatani* Marshall, *Trioxys cirsii* Curtis and *Dyscritulus planiceps* Marshall. The adults of these four wasps usually insert a single egg into the body cavity of its host sycamore aphid. Except for *D. planiceps*, which oviposits in adult aphids, they mainly oviposit in very young aphids. The larvae that hatch from these eggs complete their development within an aphid, which results in its host's death just prior to the parasitoid larva pupating. The early instar larvae of these parasitoids lack a gut and have a body surface rich in proteins associated with transepithelial transport by which nutrients are absorbed through the body wall. The later instar larvae have a gut of which the epethelial lining has a well-developed brush border. The presence of food in the gut also indicates that most food is assimilated via the gut in these instars. At oviposition the adult parasitoid injects venom, which induces a dramatic and rapid degeneration of the host aphids germarial cells and unprotected sub-apical embryos. This results in castration and as a consequence the resources previously used by the gonads of the aphid are now available for the developing parasitoid (Digilio *et al.*, 2000; de Equileor *et al.*, 2001). Thus parasitoid larvae do not have to compete with the aphids' otherwise rapidly growing gonads for resources. Interestingly, the symbionts appear to be unaffected by the venom (Tremblay & Iaccarino, 1971; Pennacchio *et al.*, 1995), and continue to upgrade and recycle nitrogen and may even furnish additional essential nutrients (Rhabé *et al.*, 2002) for the parasitoid larva's rapid growth.

Life-history strategies of the parasitoids

Aphelinus thomsoni

On locating a young aphid this wasp moves to the anterior half of the aphid, approaching very slowly, its antennae vibrating gently in front. It stops very close to the aphid, just touching it with the tips of its antennae for four to five seconds, then quickly turns through 180 degrees and inserts its ovipositor into the aphid. While doing this it places its hindmost pair of legs on top of the aphid and holds its wings vertically (Figure 4.1). The aphid initially struggles violently but soon becomes immobile while the wasp continues oviposition, which lasts on average three minutes.

Males and females are frequently observed feeding on aphid honeydew. In common with other chalcid wasps, however, *A. thomsoni* females also feed on the body fluids of its host that ooze from the puncture hole made with the ovipositor. Aphids are attacked either for oviposition or for feeding. The wasp also prefers to feed on young aphids and the approach and attack is the same as an oviposition attack. When an aphid is fed upon it is completely paralysed and does not recover. The ovipositor remains inserted for twice as long in a feeding attack

Figure 4.1. An aphelinid parasitoid parasitizing an aphid.

as in an oviposition attack, and subsequent feeding lasts eight minutes. Only some of the body fluids that ooze from the puncture are consumed. The female wasps first feed in this way when they are four days old, and this feeding is necessary if they are to mature eggs, as at emergence only the very largest females contain a few mature eggs. On average they kill and feed on one aphid for every two parasitized.

The overwintering adults emerge from hibernation around mid-May and immediately begin attacking aphids and continue to do so into early June. In the field the period from oviposition to mummification, when the mature larva sticks the now empty and dry exoskeleton of the aphid to a leaf, is 23 days. Surprisingly a wasp larva developing inside an aphid does not decrease the aphid's chances of survival compared to an unparasitized aphid. The larva pupates inside the remains of the aphid, which is referred to as a 'mummy' and in this case is a distinctive shiny black colour. The period of mummification to adult emergence lasts 34 days. Thus, the total development from oviposition to adult emergence takes 57 days, which is twice that required by the sycamore aphid to complete its development under the same conditions. As this wasp overwinters as an adult it has to complete its development before leaf fall. Therefore, wasps that continue ovipositing after mid-August are unlikely to complete their development before leaf fall (p. 67). Ground predators destroy most of the mummies that are dislodged from the leaves, or fall to the ground still attached to leaves.

Monoctonus pseudoplatani

On making contact with a potential host this wasp places the tips of its antennae on the aphid and brings its abdomen downwards and

Figure 4.2. A species of *Monoctonus* parasitizing an aphid.

forwards between its legs. As a result the ovipositor reaches well in front of its head and makes contact with the aphid, into which it inserts an egg (Figure 4.2). This takes about one second. This wasp also mainly oviposits in first instar aphids. The period from oviposition to adult emergence in the field lasts approximately 40 days, and the mummy is either a light or a dark brown colour. As in the previous species, the mummies are always attached to sycamore leaves.

This wasp overwinters as a larva in dark brown mummies. The colour is a consequence of the cocoon inside the mummy being more robust than that formed by the larvae in the light brown mummies. The proportion of the mummies that are dark brown in colour increases from August onwards. As with the previous wasp the date of leaf fall determines the date of last oviposition if the wasp is to complete its development to mummification before leaf fall. This date is early September for diapausing mummies and early August for the last light brown non-diapausing mummies. A few wasps continue to oviposit after these dates but their larvae are unlikely to complete their development unless they are in aphids on trees that shed their leaves late or their mothers are successful in locating and ovipositing in aphids on such trees.

The diapausing larvae pupate the following April and the adults emerge in mid to late May, with the males emerging first. Females have between 100 and 350 mature eggs in their ovaries at emergence and will mature other eggs. The number of eggs they lay is dependent on their longevity. As adults feed on honeydew they are unlikely to be food limited especially when aphids are abundant.

Trioxys cirsii

This wasp approaches aphids in the same way as *M. pseudoplatani*; however, contact with aphids is prolonged as the parasite grips the aphid during oviposition. At the apex of the abdomen is a downward-curved

Figure 4.3. A species of *Trioxys* parasitizing an aphid.

ovipositor sheath, which is opposed by two ventral and slightly upward curved prongs. The prongs grasp the upper surface of the aphid, while the ovipositor sheath grips the lower surface, and the aphid is held in this way for 4–10 seconds (Figure 4.3).

Like the previous two wasps, *T. cirsii* also mainly oviposits in first instar aphids. In the field it takes approximately 100 days for this species to complete its development from oviposition to adult emergence. Like *M. pseudoplatani* it also produces diapausing mummies in which the wasp overwinters as a larva. It is difficult to distinguish the mummies of these two wasps in the field. However, aphids containing mature larvae of this wasp tend to wander off the leaves and enter crevices in the bark of trees. As a consequence most mummies are not to be found on the leaves. Therefore, percentage parasitism based on the numbers of mummies on leaves is likely to be a gross underestimate.

Dyscritulus planiceps

This parasitoid does not parasitize the immature stages of the aphid but captures and subdues adult sycamore aphids. As a consequence its mummies are most abundant in early autumn several weeks after the summer peak in adult sycamore aphid abundance (Figure 4.4). This parasitoid is able to parasitize an adult aphid by quickly running at it and seizing it with its forelegs, which causes the aphid to drop off the plant. After a struggle the parasitoid inserts its ovipositor ventrally into

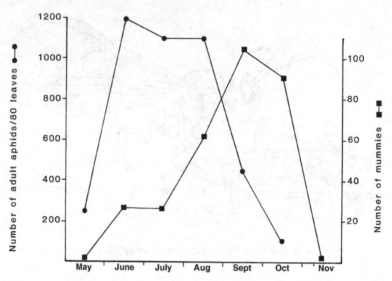

Figure 4.4. Seasonal trend in the total number of *Dyscritulus planiceps* mummies recorded in 40 tree-years relative to the trend in adult sycamore aphid numbers on one tree.

the abdomen of the aphid. Success depends on surprise, and there is little risk as parasitoid and parasitized aphid can easily fly back up and resettle on a leaf. This strategy is referred to as 'sacking' as it resembles 'sacking the quarterback' in North American football (Völkl & Mackauer, 1996).

Unlike the previous two wasps, the larvae of this species spins a cocoon beneath the aphid's body. It is not unlike the cocoon of *Praon* spp., but differs in that it is much flatter than the tent-like cocoons of *Praon*.

Endaphis perfidus

Although not recorded in this study there is a gall midge, which has been recorded in England, that parasitizes the sycamore aphid. Most of the gall midges associated with aphids are predators and a few are endoparasitic. Over 100 years ago Kieffer described *E. perfidus*, a solitary endoparasite of the sycamore aphid in Europe (Kieffer, 1896). If it is like *E. maculans* then *E. perfidus* does not oviposit its eggs into an aphid, but lays them next to but never on the body of an aphid, and upon hatching the motile larva searches for an aphid into which it burrows. On

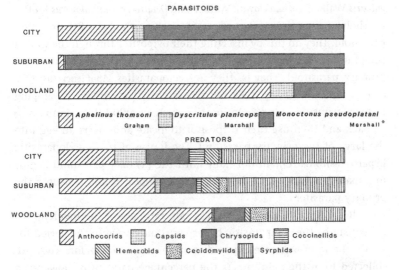

Figure 4.5. The species composition of the parasitoid and predator guilds associated with the sycamore aphid at the city, suburban and woodland sites. (*Trioxys cirsii form similar mummies, which may have been mistaken for those of M. pseudoplatani, but as the former mainly pupate off the leaves it is unlikely to have been many.)

reaching maturity the larva leaves its host through its anus and pupates in the soil (Tang et al., 1994).

Parasitoid load

Although the total number of mummies relative to the total number of aphids – i.e. the parasitoid load – recorded on the trees at the three sites (p. 20) was the same, the percentage of this total made up of the three species A. thomsoni, D. planiceps and M. pseudoplatani differs (Figure 4.5). M. pseudoplatani was the most abundant at the city and suburban sites, especially the latter, and A. thomsoni was the most abundant at the woodland site. Dyscritulus planiceps was the least abundant at all sites in Scotland, especially the suburban site.

Hyperparasitoids

The specific primary parasitoids of the sycamore aphid are attacked by generalist hyperparasitoids. Of those that were reared from sycamore aphid mummies collected at the three study sites in Scotland, *Asaphes vulgaris* Walker, *Coruna clavata* Walker and *Dendrocerus aphidovorus* Kieffer lay their eggs on the surface of mature larvae or pupae of the primary parasitoid. They do this by inserting their ovipositor through the wall of the mummy. The larvae that hatch from these eggs then consume the primary parasitoid. That is, they are ectoparasites. *Aphidencyrtus aphidovorus* and *Alloxysta* sp. seek out live aphids that contain developing primary parasitoid larvae. On locating such an aphid they climb on to its back and then use their ovipositor to locate and insert an egg into the larva of the primary parasitoid. The larvae of these endoparasitic hyperparasitoids remain dormant until the primary parasitoid begins to pupate at which stage they resume development and consume the primary parasitoid.

It is difficult to obtain an accurate estimate of the incidence of parasitism by the ectoparasitic hyperparasitoids because the period for which the mummies are exposed to hyperparasitoids before they are collected from the field affects the percentage parasitism. However, a greater percentage of the mummies, of both *A. thomsoni*, and *M. pseudoplatani*, which develop late in the season, give rise to hyperparasitoids. In *A. thomsoni*, 11% of the first generation mummies and 22% of the second generation gave rise to hyperparasitoids. In *M. pseudoplatani* it was 15–48% of the non-diapausing mummies and 44–65% of the diapausing mummies. In addition, the species of hyperparasitoid recorded from mummies differ between the sites, with the city site equally dominated by *Asaphes vulgaris* and *Aphidencyrtus aphidovorus*, and the woodland site by all four species (Figure 4.6).

PREDATORS

Predatory insects are free-living in both the adult and larval stages. They pursue, subdue and kill their prey immediately on capture, and need to consume a number of prey individuals to provide sufficient food to complete their development. Predators are usually larger than their prey, and the food source of adults and immatures is often the same. Numerically, anthocorids and syrphids were the most abundant insect predators at all three sites, closely followed by chrysopids. In addition, there were capsids, cecidomyiids, coccinellids and hemerobids (Figure 4.5).

Anthocorids were recorded most frequently at the woodland site. Of the two species, *Anthocoris nemorum* (Linnaeus) was the most abundant at the woodland site and *A. confusus* Reuter at the city and suburban sites (Figure 4.7). The adults emerged from hibernation in spring

HYPERPARASITOIDS

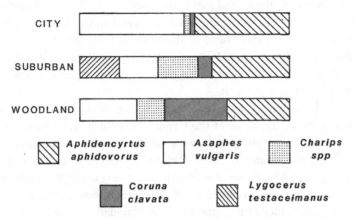

Figure 4.6. The species composition of the hyperparasitoids parasitizing the parasitoids at the city, suburban and woodland sites.

ANTHOCORIDS

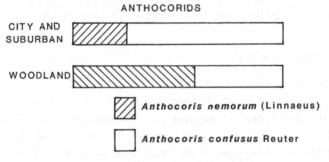

Figure 4.7. The proportion of the two species of anthocorids at the city and suburban, and woodland sites.

and were seen laying eggs from late April to late June. The nymphs that hatch from these eggs feed mainly on aphids and are dependent for their initial survival on the abundance of young aphids present at that time. Survival of the latter nymphal stages, which tend to be present in July when there are very few young aphids, is dependent upon the presence of parasitized aphids – the more mummies there are, the better their survival (Dixon & Russel, 1972). They only complete one generation on sycamore and mainly feed on the aphids present in the first half of each year.

Syrphids were also frequently recorded at the woodland site. The most abundant species were *Episyrphus balteatus* (DeGeer) and

Syrphus vitripennis Meigen. Their larvae were most frequently observed on sycamore in the second half of the year. As with the anthocorids the survival of the young syrphid larvae is dependent on the availability of young aphids. In most years at the woodland site the spring peak in abundance of young aphids was considerably smaller than the autumn peak, and this possibly accounts for why so few syrphid larvae were observed early in the year.

The larvae of the lacewings belonging to the Chrysopidae and Hemerobiidae, like those of the syrphids, were also mainly observed in the second half of the year. This was particularly so for the hemerobids.

Of the remaining predators the black-kneed capsid *Blephari-dopterus angulatus* Fallen was mainly, and the two-spot ladybird *Adalia bipunctata* (L.) only, recorded at the city site. In both cases the immature stages exploited the spring peak in abundance of young sycamore aphids. There is only a very short period in late June in most years when young aphids are sufficiently numerous for the survival of the ladybird larvae (Dixon, 1970a). In contrast, the larvae of the predatory cecidomyiid *Aphidoletes aphidomyza* (Rondani) were clearly most abundant at the woodland site, and only present late in the year.

Predator load

Unlike the parasitoid load, the total predator load was four times greater at the woodland site than at either the city or suburban sites. This is mainly because of the greater absolute and relative abundance of anthocorids and syrphids at the woodland site. In addition, the insect predators are either mainly associated with the spring peak in abundance of young aphids, i.e. anthocorids, capsids and ladybirds, or the autumnal peak, i.e. cecidomyiids, chrysopids, hemerobids and syrphids. At the woodland site the average autumnal peak in abundance of young aphids was greater than that of the spring peak. Therefore, it is likely that the incidence of larvae of cecidomyiids, chrysopids, hemerobids and syrphids on sycamore is even more constrained by the abundance of young aphids than are the immature stages of the other predators.

HABITAT QUALITY

The species composition of the natural enemies at the three sites varied greatly. Assessing their impact on aphid abundance, especially that of the insect predators, which vary greatly in size, is difficult in the absence of a detailed knowledge of how many aphid individuals each of the natural enemies kills. However, if one ignores the differences in size and treats all the species of parasitoid and all the species of predator as similar in terms of their impact then a pattern emerges. The three habitats are very similar in terms of parasitoid load, but the aphids at the woodland site are subject to a considerably heavier predator load than at the other two sites.

In summary, the guild of natural enemies attacking the sycamore aphid are all, with the notable exception of *D. planiceps*, dependent initially on the availability of very young aphids. Therefore, the seasonality of sycamore aphid population dynamics and the time of leaf fall are likely to have been important in determining the life-history strategies of the natural enemies, particularly the specific parasitoid wasps. Even the parasitoids have relatively long life cycles (40–100 days) compared to the aphid (28 days) and as a consequence are unable to track aphid abundance closely, especially as they are also constrained by the time of leaf fall. In addition, the parasitoids are heavily parasitized by hyperparasitoids, especially in autumn, and the mummies are frequently the prey of predators. The species composition of the parasitoid guild like that of the predator guild differed between sites. Anthocorids were found at all sites but the larvae of cecidomyiids, chrysopids, hemerobids and syrphids were mainly associated with the autumnal peak in abundance of young aphids at the woodland site. Although in terms of parasitoid load the three sites are similar the woodland site has a considerably heavier predator load than either the city or suburban sites.

5

Carrying capacity of trees

The sigmoidal shape of the growth of human populations was first represented mathematically by Verhulst (1838) and subsequently independently by Pearl and Reed (1920). This led to attempts to use laboratory populations to verify what came to be referred to as the logistic model. Although this is not the only mathematical model that fits population growth curves it has the great merit of being simple and biologically realistic. It neatly encapsulates the concept that populations increase until the demands made on the resources preclude further growth and the population is then at its saturation level, a value determined by the carrying capacity of the environment (Varley *et al.*, 1973; Begon & Mortimor, 1981; Berryman, 1981; Renshaw, 1991). Saturation level/point was first used by Lotka (1924), and carrying capacity first coined by Errington (1934) and Errington and Hamerstrom (1936). Initially carrying capacity was used to specify the maximum number of bobwhite quail an environment could sustain. Later Errington (1946) used it when referring to populations of other vertebrates, especially birds and rodents, which showed some form of territorial behaviour. Carrying capacity as a concept proved very attractive and was quickly equated with K in the logistic model and replaced saturation point and level. In 1988 carrying capacity was voted the seventeenth most popular ecological concept by members of the British Ecological Society (Cherrett, 1988).

Generally, the carrying capacity of an environment for a particular species is thought to remain constant in time, although some authors indicate it can increase or decrease (Odum, 1953), or is affected by weather (Dempster, 1975) or genetic changes, which affect the efficiency with which individuals utilize resources (Ricklefs, 1990). The logistic model also gave rise to the concepts of opportunistic and equilibrium populations. Individuals in populations kept below the carrying capacity of the environment by periodic disturbances are likely to be selected for high rates of increase (r selected). Those in populations that are at or close to carrying capacity are likely to be selected for competitive ability rather than a high rate of increase (K selected) (MacArthur & Wilson, 1967). This very appealing idea was eagerly embraced by many ecologists and generated a lot of literature. Of particular relevance to what follows is that the carrying capacity of long-lasting habitats (woodlands) is thought to be fairly constant (Southwood, 1976). However, when a herbivore – such as a deer, which eats its limiting resource – becomes abundant it might result in overgrazing and an absolute shortage of food, which results in a dramatic and long-lasting reduction in the carrying capacity (Leopold, 1943). This does not apply to the sycamore aphid–host interaction but possibly is relevant to other aphid tree systems (p. 118).

In summary, logistic theory and field and laboratory studies gave rise to the idea that each environment can sustain only a certain number of a particular species, which is referred to as the carrying capacity for that species. Although a very attractive idea it is difficult to define, especially in the case of insects. An attempt to define it for the sycamore aphid follows.

APHID POPULATION DYNAMICS AND THE CONCEPT OF CARRYING CAPACITY

The average number of sycamore aphids per leaf varies dramatically both spatially and temporally. However, both show a characteristic seasonal trend. In terms of vertical distribution there are few or no aphids in the upper canopy of mature trees in early summer, and the population density is generally at its lowest on sycamore trees in late summer. Even when population density is very high there are often many leaves upon which there are few or no aphids. This was attributed to self-imposed aggregation by Kennedy & Crawley (1967), but is more likely a consequence of spatial heterogeneity in microclimate, with some leaves more favourable than others (Dixon & McKay, 1970) (p. 35). Similarly, the vertical movement within the canopy, with virtually all the aphids in the lower canopy during summer (see Figure 3.8), is also most likely a response to seasonal changes in microclimate. In the summer the microclimate of the leaves in the upper canopy is possibly less favourable for aphids than that of the leaves in the lower canopy (p. 29). One consistent theme in this is that the favourableness or otherwise of the habitat is measured in numbers of aphids per leaf.

CARRYING CAPACITY OF LEAVES

In the laboratory

The nutritive quality has a marked effect on how long an aphid will remain feeding on a leaf. An aphid that settles on a mature leaf is five times more likely to move off per unit time than an aphid on a young or senescent leaf. Movement is also more marked at higher temperatures. An increase of 8 °C approximately doubles the incidence of movement. That is, even if an aphid is on its own it will sooner or later walk off a leaf on which it has settled (Dixon, 1970b). This will occur sooner if the leaf is of poor quality and the temperature is high than when the leaf is of high quality and the temperature is low.

In early spring and late autumn, sycamore aphids are often seen in tightly packed groups, possibly because both the high nutritive status of the host and low temperature reduce the restlessness of aphids. In summer, when aphids are abundant, the nutritive quality of the host is poor and temperatures are higher, they frequently move from leaf to leaf (pp. 26–8), which increases the likelihood of encounters between aphids (Dixon, 1970b). That is, in addition to the physical limit imposed by leaf area, the tendency of aphids to move from leaf to leaf – i.e. their restlessness – is implicated in determining the carrying capacity.

In an attempt to determine whether leaf area influences the number of aphids that settle on a leaf, third instar sycamore aphid nymphs were placed individually, over a period of an hour each day, on the petioles of similar sized leaves of sycamore saplings. The number of aphids placed on the petioles each day was 10 or 20 and this was continued for 11 days. The results are illustrated in Figure 5.1. In both cases the number of aphids on the leaves increases, but stabilizes at different levels. The levels at which the number stabilize is clearly related to the number of aphids placed on the petiole each day, and even when it is 20 aphids per day not all the leaf area is occupied by aphids. In addition to those that walked onto a leaf each 24 hours there are others that must have walked off. That is, the carrying capacity of a leaf, in this situation, was achieved when the number of aphids walking on to a leaf equalled those walking off every 24 hours. When 20 aphids were placed on petioles of similar sized leaves of saplings kept at a range of temperatures, the apparent carrying capacities of these leaves declined with an increase in temperature (Figure 5.2). This indicates that leaf area appears to be relatively unimportant in determining the carrying

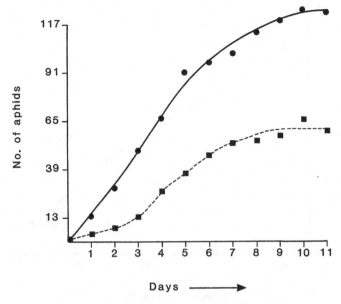

Figure 5.1. The trends in the numbers of sycamore aphids that settled on two similar sized leaves of sycamore saplings kept at 17 °C when 20 (•——•) or 10 (■ – – – ■) third instar aphids were placed individually on the petioles of the leaves over a period of one hour each day on 11 consecutive days.

capacity of leaves for sycamore aphid nymphs, at least at the invasion rates used.

In the field

Clearing leaves of aphids each day in the field indicates that aphids are continually moving between leaves (pp. 26–8). The carrying capacity of leaves in the field was determined over a short period in summer when populations consist entirely of adults and the numbers remain relatively constant. Prior to the start of the experiment aphids were counted and cleared daily from seven leaves. Starting on 21 June, the aphids were counted but not removed from two of the leaves. From the relationship between the number of aphids moving on to the other five leaves and these leaves, derived over the previous 10 days, the number moving on to the two experimental leaves could be estimated. The results are illustrated in Figure 5.3. The numbers increased and then stabilized, and similarly after 6 July when the leaves were cleared again, the number

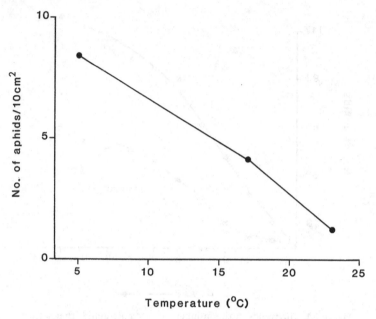

Figure 5.2. The average number of aphids per unit area that settled on similar sized leaves of sycamore saplings kept at a range of temperatures. Each day, 20 third instar aphids were placed individually over a period of an hour on to the petioles of the leaves. The average number was the average of the numbers recorded as settled on the leaves over a period of six days once the numbers had ceased to increase.

of aphids increased and stabilized. The levels at which they stabilize on a particular leaf in each period is similar, and in terms of the number of aphids per unit area is similar on both leaves. That is, as in the laboratory, the maximum number that settle on a leaf is less than the number estimated to have walked on to the leaf during the course of the experiment. However, in this case the carrying capacity does appear to be associated with leaf area. That is, when the aphids are in reproductive diapause and at the time of the peak number of aphids in the lower canopy, the number on suitable leaves appears to be at or close to the physical carrying capacity.

Although gregarious, the sycamore aphid is more widely spaced at low than at high overall population densities (p. 81). That is, on those leaves with the highest population densities, which are likely to be the most favourable microclimatically, the aphids are most closely packed and competing for favourable space. In spacing themselves out in this way mutual stimulation at low population densities is reduced and

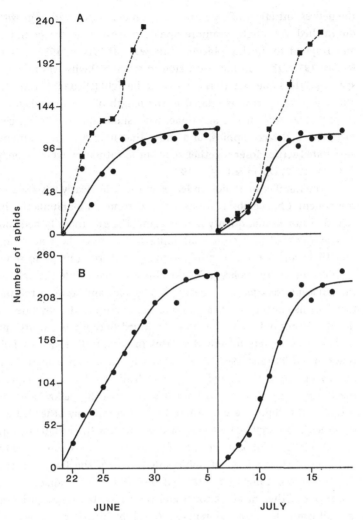

Figure 5.3. The trends in the numbers of adult aphids settled on two leaves in the field that were cleared of aphids on 21 June and 6 July. Leaf A had an area of 75 cm with a maximum of 1.6 aphids/cm and that of leaf B 122 cm and 1.9 aphids/cm. The cumulative daily recruitment on to leaf A (■ – – – ■) was estimated from the relationship between number of adults moving on to five other adjacent leaves and leaf A over the previous 10 days.

the depressant effect of gregariousness on aphid population growth is diminished. Although sycamore aphids possess large compound eyes and respond to moving objects some way off (Dixon, 1958; Dixon & McKay, 1970), they do not use vision in spacing themselves out as the spacing is the same in both the dark and the light (Dixon & Logan, 1972). The degree of spacing is related to the length of the appendages and in particular that of the antennae, with small aphids settling closer together than large aphids, and those that have had their antennae amputated settle closer together than those with antennae (Kennedy & Crawley, 1967; Dixon & Logan, 1972).

The number of aphids on leaves in the field clearly has a seasonal component. Chance observations and experimental manipulation indicate that nutritional quality is important. The guy ropes lashed round a sycamore tree to keep a scaffolding tower steady became so tight that they caused the leaves in the top part of the canopy to senesce in midsummer. The consequence of this was that, while the aphids in the adjacent trees vacated the upper canopies and were in reproductive diapause in the lower canopies of these trees, in the strangled tree the aphids not only remained but continued to reproduce in its upper canopy. All attempts to reproduce this in a controlled fashion failed. However, on 29 June 1962, it was noticed that a branch in the lower canopy of one of the sample trees was badly cracked at its base. On the 19 July the leaves on this branch were noticeably yellow, and more aphids and nymphs were on these leaves than on any other leaves of the tree. Occasionally the leaves of twigs within the canopy of a tree senesce while those on adjacent twigs do not. In 1964, from mid-June to mid-August, the level of aphid infestation of the leaves on seven such twigs was compared with that on the leaves of adjacent twigs. The leaves of the cracked branch and the twigs that senesced in summer all contained proportionally more soluble nitrogen and less total nitrogen than adjacent mature leaves, which indicates that nitrogen had been translocated out of these leaves (Figure 5.4). This possibly accounts for the greater number of adult aphids and reproductive rate of the aphids on the senescent leaves (Figure 5.5). Similarly, petioles are never colonized by aphids in summer. However, several days after removing the lamina from such leaves the petioles senesce and are colonized by aphids, which may even begin reproducing before the petioles are finally shed (Dixon, 1976b). That is, the carrying capacity of leaves during summer is very markedly affected by their nutritive status, and this is most striking in the case of the leaves in the upper canopy.

NITROGEN

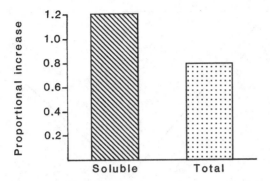

Figure 5.4. The proportional increase in quantity of soluble and total
nitrogen in senescent compared with mature sycamore leaves in
summer. (After Dixon, 1966.)

APHIDS

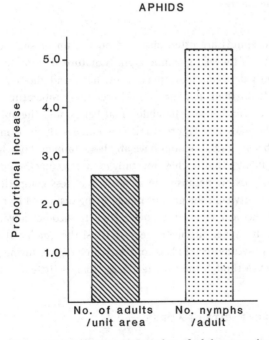

Figure 5.5. The increase in number of adults per unit area of leaf and
number of nymphs per adult on senescent relative to mature sycamore
leaves in summer. (After Dixon, 1966.)

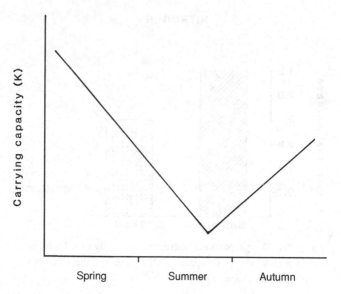

Figure 5.6. Seasonal trend in the sycamore aphid carrying capacity of a mature sycamore tree.

In autumn, aphids are often observed on the upper surface of leaves. At this time of the year sycamore aphids mature, due to the low temperatures, with dark-coloured appendages, head and thorax, and eight black bands on their abdomen. The lower the temperature the more extensive and blacker are the abdominal bands, and the aphids can become quite black (see Figure 3.11). This enables them to more effectively absorb solar radiation and have an above ambient body temperature (Dixon, 1972b). That is, they normally do not occupy the upper surface of leaves, possibly because the advantage of less competition for space does not outweigh the greater risk of being dislodged by rain. However, late in the year when it is cooler then it becomes advantageous to occupy the upper surface of leaves. Thus, the quality of the habitat available to sycamore aphids is continually changing during the course of a year even though the total leaf area changes little if at all.

CARRYING CAPACITY OF A TREE

The carrying capacity of a mature tree reflects that of individual leaves. In spring, leaves at all levels in the canopy are suitable for colonization by sycamore aphids. In summer, when leaves generally are of poor nutritive quality, those in the upper canopy become completely unsuitable because of the higher temperatures they experience. At this time the aphids go into reproductive diapause, vacate the upper canopy and aggregate in the lower canopy. In late summer–early autumn, the nutritive quality of the leaves begins to improve, temperatures decline and the aphids begin to recolonize the middle and upper canopy. This coincides with a marked exodus of aphids from mature trees, and the lowest overall aphid abundance. The numbers on mature trees in autumn increase generally throughout the canopy as the leaves senesce prior to leaf fall. That is, the carrying capacity of mature sycamore trees for the sycamore aphid changes dramatically during the course of a year (Figure 5.6). It starts high in spring when the leaves are growing, reaches a low level in late summer when the leaves are mature and recovers again in autumn when the leaves senesce. This has important consequences for sycamore aphid population dynamics (pp. 114–16).

In summary, the carrying capacity of a tree is determined by spatial and temporal changes in the microclimate and nutritive quality of leaves. In spring sycamore trees generally are highly suitable for aphids, while in summer only some of the leaves in the lower canopies are suitable, and with the approach of autumn they generally become gradually more suitable, especially those in the upper parts of the canopy. That is, there is a marked seasonal trend in carrying capacity, which is high in spring and autumn and low in late summer. This is closely associated with seasonal changes in the nutritional quality of the host plant and the reproductive activity of the aphid.

6

Aphid abundance

In Chapters 3–5, sycamore trees as a habitat for aphids are considered. This chapter considers the factors that are likely to determine the level of abundance at which sycamore aphid populations are regulated. That is, the non-reactive or 'purely density-legislative' and regulative (density dependent) factors (Nicholson, 1954). In addressing this topic the role of the abundance and phenology of sycamore, the size of the aphid, aphid-induced changes in the quality of sycamore, natural enemies and intraspecific competition are considered. All of which could affect aphid abundance.

ROLE OF HOST ABUNDANCE

The equilibrium population density of a species is seen as the outcome of the interaction between its rate of increase and density dependent mortality factors. Natural enemies do not appear to regulate the abundance of tree-dwelling aphids (p. 75). Moreover, there is no evidence that the efficiency of the natural enemies of the different species of aphid differ, therefore, differences between aphid species in their intrinsic rate of population increase, r_m, or more particularly realized rate of increase, R, appear to be the most likely cause of differences in abundance between species. The theoretical grounds for this are presented in Dixon and Kindlmann (1990).

The between-year dynamics of tree-dwelling aphids can be represented by:

$$\log X_{t+1} = \log X_t + \log R - M \log X_t \qquad (6.1)$$

Where X_t and X_{t+1} are the peak numbers in spring in years t and $t+1$, R the realized rate of increase and M the density-dependent factor. After antilogging, equation (6.1) gives:

$$X_{t+1} = R X_t^{1-M} \qquad (6.2)$$

The equilibrium density (X^*) of which is:

$$X^* = R^{1/M} \qquad (6.3)$$

What factors are likely to affect the degree to which r_m is realized? Dixon et al. (1987) argue that one such factor is the probability of finding a host plant. This is likely to be directly proportional to the ground covered by the host plant $[P(C)]$ assuming aphids disperse regularly. Thus, the realized rate of increase, R, which includes the losses incurred in dispersal is:

$$R = r_m P(C) \qquad (6.4)$$

Given that the probability of finding a host plant $[P(C)]$ after D trials is:

$$P(C) = 1 - (1 - C)^D \qquad (6.5)$$

Then the equilibrium density is given by:

$$X^* = \{r_m [1 - (1 - C)^D]\}^{1/M} \qquad (6.6)$$

This indicates that all other things being equal, the proportional cover of the host plant, through its effect on realized r_m, can markedly affect the abundance of an aphid.

The indigenous deciduous tree-dwelling aphids of Britain all belong to the same subfamily, Drepanosiphinae. They are either highly host-specific or live on at most two species of a particular genus of tree. The relationship between ranked abundance of these aphids and that of their host plants is presented in Figure 6.1. This lends support to the idea that plant abundance is a major factor determining aphid abundance.

ROLE OF HOST PHENOLOGY

There has been a tendency to attribute differences in the abundance of aphids on trees to their resistance to aphids or exposure to wind. The lime aphid, which lives on several species of lime tree, is often more abundant on the small- than on the large-leafed lime trees growing on the same site (Figure 6.2). It is likely this is due to differences in the trees, possibly their nutritive quality. In the case of the eight sycamore trees studied, aphids were consistently very abundant on some trees and less abundant on others over long periods of time (Figure 6.3).

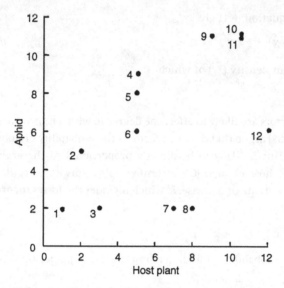

Figure 6.1. The relationship between rank order of abundance of 12 species of deciduous tree-dwelling aphids and the rank order of abundance of their host plants ($r_s = 0.52$, $P < 0.05$).

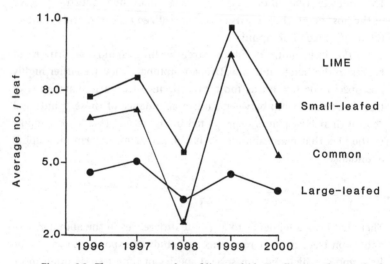

Figure 6.2. The average number of lime aphids *Eucallipterus tiliae* per leaf from May to September each year from 1996–2000 on small (*Tilia cordata*), common (*T. x europea*) and large-leafed (*T. platyphyllos*) limes growing in the same area. (After Juronis, 2001.)

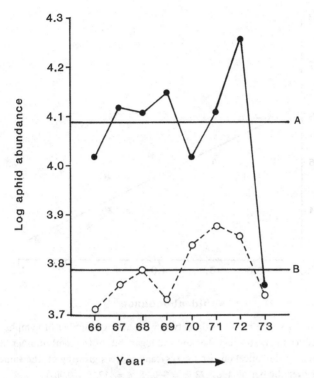

Figure 6.3. The logarithm of the total number of sycamore aphids counted each year from 1966–73 on two trees growing on the same site. The average abundance on tree A was twice that on tree B.

For the two trees depicted, the average abundance on the more heavily infested tree was twice that on the less heavily infested tree in spite of the reproductive rate of the aphids on the lightly infested tree being greater than those on the heavily infested tree (Figure 6.4). It is likely that the difference in the abundance on the two trees is due to marked differences in phenology of the trees. The more heavily infested tree on average sheds its leaves 17 days later and breaks its buds 11 days earlier than the less heavily infested tree (Figure 6.5). The relationships over a period of 10 years between (a) the numbers of aphids on the leaves in autumn and those on the buds the following spring, and (b) the numbers on the buds and colonizing the leaves in spring, indicate that proportionally more aphids survived these two events on the more heavily infested tree. It was 0.42 and 0.61 on the more heavily infested tree, and 0.25 and 0.12 on the less heavily infested tree ($t = 2.3$, d.f. $= 10$, $P < 0.05$; $t = 3.3$, d.f. $= 10$, $P < 0.01$, respectively). For two trees on

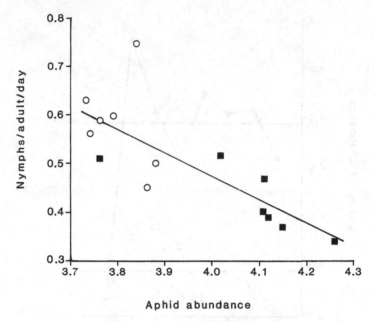

Figure 6.4. The relationship between the average number of nymphs produced per adult per day and the logarithm of the total number of aphids (abundance) on two (o, ■) sycamore trees growing on the same site over the period 1967–73 ($r = -0.78$, $n = 12$, $P < 0.001$).

another site, which did not differ in the average times of leaf fall and bud burst, the proportions of aphids that survive these two bottle-necks did not differ ($t = 0.9$, d.f. $= 10$, NS; $t = 1.03$, d.f. $= 10$, NS, respectively).

For the eight trees there is a significant rank correlation between average aphid abundance and the average proportion overwintering ($P = 0.05$). In addition, the variation in the relationship between spring and autumn numbers for each of the trees is also similarly significantly correlated with the average proportion successfully overwintering on each tree ($P = 0.05$).

Thus for the eight trees spread over three sites there is strong evidence to indicate that a major determinant of aphid abundance is the overwintering mortality suffered on each tree. This is closely associ-ated with the degree of synchrony between egg-laying and leaf fall, and egg hatch and bud burst. That is, synchronization between aphid and tree phenology is a key determinant of the difference in abundance of aphids between trees. Another factor, which may account for some of the variation in the above relationships, is the extent to which a tree

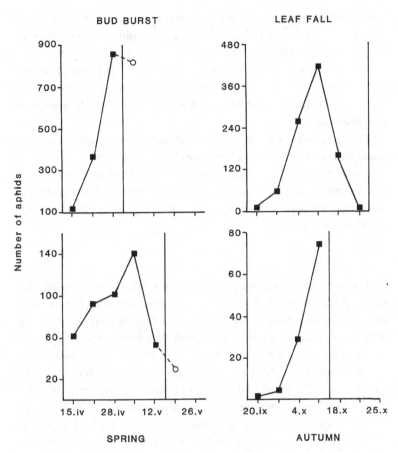

Figure 6.5. Trends in aphid abundance at bud burst and leaf fall on heavily infested (upper) and lightly infested (lower) sycamore trees growing on the same site in 1969. The vertical lines represent the times of bud burst and leaf fall, respectively.

is isolated from other trees. Isolated trees may support fewer aphids because they lose more aphids than they gain by dispersal than trees growing in clumps (Dixon, 1990b).

In the case of the Turkey oak aphid system the two trees sampled were similar in terms of the success of the aphid switching from asexual to sexual reproduction and survival of the overwintering eggs. However, the leaves of the more heavily infested tree senesced earlier and the leaves were shed some two weeks earlier than those on the other tree. As a consequence, although the reproductive rate in the first half of a year is the same on both trees, it is very much higher on the heavily infested tree in the second half of the year. The difference in average abundance on the two trees (x1.4) is similar to the difference in the average fecundity achieved on the two trees (x1.5), therefore, the difference in abundance may be accounted for in terms of recruitment. However, in this system, as in the sycamore-aphid system, the difference in well-being of the aphid is associated with differences in tree phenology.

In summary, the degree of synchrony between aphid and tree phenology is an important determinant of the differences in abundance of aphids on trees and variation in abundance from year to year on a particular tree. In addition, the degree of isolation from other trees of the same species may also be important in determining the abundance of aphids on a tree.

BODY SIZE

Gaston and Lawton (1988a,b) and Lawton (1989) have proposed that small species of insect are more abundant because they have a lower per capita resource requirement than large species. Although this may hold when comparing generalist species of a range of sizes from different taxa, there is no evidence for it at the level of a taxum. For example, both the field maple *Acer campestris* and sycamore host similar sized specific aphids that belong to the same genus (*Drepanosiphum aceris* and *D. platanoidis*), which have similar intrinsic rates of increase and are attacked by the same parasitoids and predators. However, *D. aceris* is much less common on field maple than is *D. platanoidis* on sycamore (Stroyan, 1977). That is, there is no evidence to indicate that an aphid's abundance is mainly associated with its per capita resource requirement.

APHID-INDUCED CHANGES IN THE SEASONAL TREND IN PLANT QUALITY

In general, high aphid numbers in spring are followed by low numbers in autumn and vice versa. The 'seesaw' effect in the sycamore aphid is associated with marked differences in autumnal reproduction (see pp. 84–6). Initially the between-year differences in the autumnal reproductive performance were thought to be driven by induced changes in the quality of its host. This was supported by the marked effect this aphid has on the growth and senescence of the leaves of sycamore (p. 154; Dixon, 1970b, 1971a). The idea that plants respond to herbivory by becoming less suitable and so regulate the abundance of herbivores (Haukioja & Hakala, 1975) is a popular one and difficult to ignore.

The above hypothesis was tested by recording: (a) the nitrogen content of the leaves of sycamore in the middle of a year and again at leaf fall for each of the trees sampled in the field over a period of six years; (b) the seasonal changes in the concentrations of amino acids in the leaves of a tree and those of previously infested and uninfested saplings; and (c) performance of aphids on previously infested and uninfested saplings.

Nitrogen content of the leaves

The percentage of nitrogen in the leaves in the middle of a year does not differ between trees on a particular site. However, it does differ between sites and appears to be associated with the nitrogen status of the soil. It was highest for the two trees growing in alluvial soils on what had previously been a rose nursery, the suburban site; intermediate for the woodland site; and lowest for the leaves from the two trees growing close to buildings, the city site. The recovery of nitrogen from the leaves prior to leaf fall was more variable between trees on each site and the percentage recovered appeared to be associated with the nitrogen status of the soil. The two trees on the poorest site recovered the most nitrogen; the four on the woodland site recovered an intermediate amount; and the two growing on the rich alluvial soils recovered least. More importantly there was no correlation between the amount of nitrogen recovered from the leaves prior to leaf fall and the peak numbers of aphids on the trees in autumn for each tree, site and overall.

Amino nitrogen content of the leaves

The content and percentage recovery of nitrogen from leaves prior to leaf fall gives a crude measure of their nitrogen metabolism. As a consequence it was decided to measure changes that occur in the wide range of free amino acids present in sycamore leaves (Wellings & Dixon, 1987). Studies using synthetic diets have shown that the amino acids histidine, isoleucine, lysine and methionine are essential for the growth of the aphid *Myzus persicae* (Sulzer) (Dadd & Krieger, 1968) and histidine, serine, proline, cysteine, methionine, tyrosine, phenylalanine and alanine for the growth of *Aphis fabae* Scop. (Leckstein & Llewellyn, 1973). It is possible that the changes observed in sycamore aphid reproductive activity could be determined by changes in the availability of a few essential amino acids rather than the total free amino acid content of the leaves. This was tested by determining the degree of association between the seasonal changes in aphid recruitment and changes in specific amino acids, groups of essential amino acids or the total free amino acid content of the sycamore leaves, in addition to the direct effect of aphid population density. There is no direct or delayed association between reproductive activity and any of the 24 amino acids assayed or groups of 'essential' amino acids. The changes in the abundance of essential amino acids are correlated with the total free amino acid content of the leaves (Dixon et al., 1993). The lack of a clear association between aphid reproductive activity and the concentration of specific amino acids or groups of amino acids may be because the sycamore aphid's symbionts upgrade unessential to essential amino acids (p. 15), and this aphid is therefore able to utilize most if not all the amino-nitrogen in its food.

Aphid infestation at bud burst has a marked effect on the growth of sycamore – e.g. reducing the area of the leaves to less than half that of the leaves of uninfested trees (pp. 154–5; Dixon, 1971a). However, analysis of the free amino acid composition of the leaves indicate that no major changes in leaf quality occur in autumn as a result of prolonged aphid infestation of saplings at bud burst and over the period of leaf growth. The free amino acid concentration in the leaves of sycamore that had previously been infested for 8 and 16 weeks from bud burst and in uninfested saplings is not associated with previous infestation history (Wellings & Dixon, 1987). That is, there is no evidence from the amino acid content of leaves to support the hypothesis that sycamore responds to high aphid infestation by becoming less suitable for aphids in autumn.

Aphid performance

Sycamore saplings subject to very heavy aphid infestations in spring are no less suitable as hosts for aphids in autumn than previously uninfested saplings (Dixon, 1975a). In addition, there is no evidence of an induced response, even in saplings exposed to high levels of aphid infestation in both the previous autumn and spring. That is, conditions that more closely simulate those in the field. This is the case even when aphids are stressed by being crowded. Second generation adults collected from the field and clip caged individually on the leaves of previously infested and uninfested saplings do not respond differently. The only trend is that the large individuals are more fecund than the small individuals, but this is independent of the infestation history of the saplings (Table 6.1). When reared from birth to maturity, third generation individuals do better on previously infested saplings, even when reared in groups of four (Table 6.2), which is contrary to what the theory

Table 6.1. *The average initial weights, final weights, days to onset of reproduction, fecundity, weights of offspring at birth and total weights of offspring produced by batches of 20 second-generation adult sycamore aphids collected from the field on 21 July 1980 and clip caged individually for 16 days on the leaves of sycamore saplings that had either been infested with aphids the previous autumn and spring (experimental) or kept free of aphids (control)*

	Experimental	Control	Significance
Initial wt (μg)	898.3	877.2	NS
Final wt (μg)	1139.5	1078.1	NS
Days to reproduction	9.6	9.9	NS
Fecundity	181	166	NS
Wt of offspring at birth (μg)	82.2	81.9	NS
Total weight of offspring (μg)	837.2	758.4	NS

Covariance analysis of the relationships between the number of offspring produced in 16 days and initial weight for the aphids clip-caged on previously aphid infested and uninfested saplings.

Slopes $F = 1.45$, 1/36, NS; Elevations $F = 0.002$, 1/37, NS.

Table 6.2. *The average mid-point of development, birth weight, adult weight, developmental time and relative growth rate (RGR) of third generation aphids reared individually (A) and in groups of four (B) on the leaves of sycamore saplings that had either been infested with aphids the previous autumn and spring (experimental) or kept free of aphids (control) (n = number of aphids)*

	Experimental	Control	Significance
Mid-point of development	10 August	11 August	NS
A			
n	47	55	
Birth weight (μg)	85.9	84.6	NS
Adult weight (μg)	843.0	740.9	*
Developmental time (days)	18.7	21.2	**
RGR	0.1256	0.1083	*
B			
n	44	36	
Birth weight (μg)	85	84.4	NS
Adult weight (μg)	785.9	607.9	*
Developmental time (days)	18.8	23.8	**
RGR	0.1231	0.0846	**

*, $P < 0.05$; **, $P < 0.01$.

Table 6.3. *The average weight and number of well developed embryos in adult second generation sycamore aphids collected in the field on 21 July and 2 August 1980, and the average weight, number of well developed embryos and number of offspring produced by second generation adults collected on 21 July in the field and kept in crowds (10/cage = 1.3 aphids/cm²) for 11 days on the leaves of sycamore saplings that had either been infested with aphids the previous autumn and spring (experimental) or kept free of aphids (control) (n = number of aphids)*

	Field-collected aphids			Aphids collected on 21 July and caged in groups of 10 for 11 days on sycamore saplings		
	21 July	2 August	Significance	Experimental	Control	Significance
n	40	60		100	100	
Adult wt (µg)	893.3	1007	NS	1039.4	960.2	NS
No. of well developed embryos	0.69	0.45	*	4.5	3.9	NS
No. of offspring				2.6	2.0	NS

* $P < 0.05$.

predicts. Field-collected second generation adults kept in groups of 10 per cage, which is comparable to the population densities observed in the field, emerge from reproductive diapause equally rapidly on both types of foliage.

Interestingly, the second generation adults collected in the field on the 2 August were collected on the basis of the number of well developed embryos in their gonads still in reproductive diapause, whereas those brought in from the field 11 days previously and kept individually on saplings were not (Table 6.3). The fact that the aphids on mature trees in the field were still in diapause on 2 August, whereas those on saplings were not, clearly indicates that living on mature trees in the field is not the same as living on saplings in a glasshouse. However, these experiments and the amino acid analysis of the leaves of previously infested and uninfested saplings reported by Wellings and Dixon (1987) indicate that aphid infestation of saplings, even prolonged infestation, does not result in nutritional changes that adversely affect the performance of the aphids. That is, there is no evidence of aphid-induced changes in host-plant quality affecting sycamore aphid performance in autumn. The fact that crowded aphids remain in summer reproductive diapause on mature trees in the field for longer than they do on saplings in a glasshouse may indicate that there are other stresses operating in the field, in addition to crowding.

ROLE OF NATURAL ENEMIES

There is no doubt that natural enemies can reduce the survival of aphids, occasionally dramatically, and the use of hymenopterous parasitoids and ladybirds in the biological control of aphids is claimed to have been successful. Indeed the first outstanding success in biological control was the use of a ladybird (*Rodolia cardinalis*) to control the cottony cushion scale *Icerya purchasi*, a close relative of aphids. This reinforced the belief that natural enemies regulate insect abundance. Indeed the array of insect natural enemies that attack the sycamore aphid is awesome (Table 6.4) and the four primary parasitoids are host specific.

In spite of the success of *Rodolia*, ladybirds have not proved effective biological control agents of aphids (Dixon, 2000). The reason for this is to be found in the dynamics of the aphid–ladybird interaction and its consequence for ladybird fitness (Kindlmann & Dixon, 1993). A major factor in this is that aphids develop much faster than ladybirds and as a consequence aphids can disperse and become quite scarce

Table 6.4. *The parasitoids and insect predators of the sycamore aphid*

Parasitoids	Predators
Aphelinus thomsoni[a]	Adalia bipunctata
Dyscritulus planiceps[a]	Anthocoris confusus
Monoctonus pseudoplatani[a]	Anthocoris nemorum
Trioxys cirsii[a]	Chrysopa carnea
Endaphis perfidus	Chrysopa ciliata
	Episyrphus balteatus
	Syrphus vitripennis
	Tachydromia arrogans

[a], Host specific.

before ladybird larvae can complete their development. This is important because the larvae unlike the more mobile adult ladybirds cannot seek aphids elsewhere. Natural enemies that have a developmental time similar to that of their prey are potentially capable of regulating the abundance of their prey as is the case in the *Rodolia/Icerya* interaction (Dixon, 2000). Scale insects have a developmental time similar to other insects. That is, the generation time ratio (GTR) of ladybirds and their scale insect prey is close to 1:1. This GTR concept has been incorporated into a model that accurately predicts the patterns observed in the field (Dixon & Kindlmann, 1998a; Dixon, 2000).

Of the insect predators attacking the sycamore aphid, the anthocorids and ladybirds, which mainly attack the spring peak in aphid abundance, have been studied in detail (Dixon, 1970a; Dixon & Russel, 1972). They both have developmental times considerably longer than that of the aphid. The survival of the early stages of the anthocorids and ladybirds is very dependent on an abundance of young aphids and that of the latter stages of the anthocorids on an abundance of parasitized (mummies) and adult aphids. However, the proportion of the sycamore aphid population killed by both the anthocorids and ladybirds decreases as the sycamore aphid population increases in abundance. Another important predator at all the sites are syrphids (p. 50). Although they have not been studied in detail they also have relatively long developmental times and mainly attack the autumnal peak in aphid abundance, and as with the previous two predators there is no indication that they act in a density dependent way. That is, the predators do not appear to regulate sycamore aphid abundance. However, the greater predator load (p. 52) at the woodland site might contribute to the generally lower aphid population densities recorded on the trees at that site.

Hymenopterous parasitoids, which mature on one aphid and generally are thought to take approximately the same length of time to complete their development, would appear to be potentially capable of regulating aphid abundance. However, as indicated on page 53 the parasitoids in this case take considerably longer than the aphid to complete their development. Generally there are two peaks each year in the numbers of mummies of *Aphelinus thomsoni* and *Monoctonus pseudoplatani*, one in June–July and the other in August–September. As the time it takes for these parasitoids to develop from oviposition to mummification is known (pp. 45–46) it is possible to relate the number of mummies appearing to the number of young aphids present at the time of oviposition. The proportion of young aphids parasitized declines with the increase in aphid abundance both early and late in a year, with the levels of parasitizism late in a year much lower than early in a year. Furuta (2003) records a similar relationship in a 13-year study of the percentage parasitism by *Aphidius* sp. of another maple aphid, *Periphyllus californiensis*.

In the sycamore aphid system there is often no second peak of mummies even though the autumnal peak of aphids is similar in size in such years to those years when there is a second peak of mummies (Figures 6.6 and 6.7). There is a very strong association between the size of the second peak of mummies of both parasitoids and the availability of young aphids in mid-July to mid-August ($r = 0.8$, $P < 0.001$ in both cases; see Figures 6.6 and 6.7). If the summer reproductive diapause of the sycamore aphid is prolonged then there are few or no young aphids during this period and no second peak of mummies, if the diapause is short there are many young aphids and a large second peak of mummies. That is, the size of the second peak of mummies is not related to the autumnal abundance of the aphid but to the presence of young aphids in late summer. This is because the parasitoids have to complete their development before leaf fall, and ovipositing after mid-August leaves insufficient time for them to complete their development. In addition, the spring peak of mummies is not correlated with the peak number of mummies in the previous June–July or August–September. That is, not only are the parasitoids not acting as density dependent regulating factors but the dynamics of the aphid and parasitoids frequently become uncoupled in autumn.

The effectiveness of the parasitoids is also reduced by the action of hyperparasitoids and predators, which in many cases are less specific than the primary parasitoids (Dixon & Russel, 1972; Hamilton,

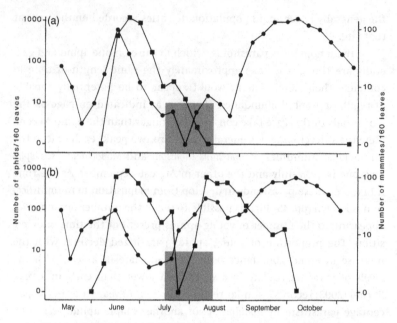

Figure 6.6. The seasonal trend in the number of young aphids (●—●) and number of mummies (■—■) of *Aphelinus thomsoni* at the woodland site in a year when there was no second peak of mummies (a) and when there was a second peak of mummies (b). The shaded block is the period during which young aphids should be abundant if there is to be a second peak.

1973, 1974; Holler *et al.*, 1993; Mackauer & Völkl, 1993). In addition, because of the risk of hyperparasitism, primary parasitoids are likely to cease ovipositing in a patch of aphids before all the aphids are parasitized. High percentage primary parasitism makes a patch attractive to hyperparasitoids. By continuing to oviposit in patches of aphids where the incidence of parasitism is high, a primary parasitoid may reduce its potential fitness (Ayal & Green, 1993).

In an attempt to reduce the abundance of the lime aphid, hymenopterous parasitoid species – *Trioxys curvicaudus* Mackauer, an *Aphidius* sp. and *Aphelinus subfavescens* (Westwood) – were introduced in 1970 from Europe into America (Olkowski, 1973), where the aphid is regarded as a social nuisance because of the large quantities of honeydew it produces. Only *T. curvicaudus* became established and initially was claimed to be effective (Olkowski *et al.*, 1982; Zuparko, 1983). After a seven-year study, however, Dahlsten *et al.* (1999) concluded that there is no evidence to indicate that *T. curvicaudus* regulates the

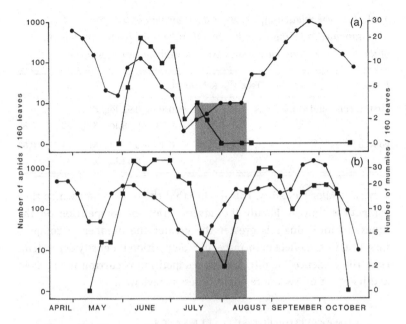

Figure 6.7. The seasonal trend in the number of young aphids (●—●) and number of mummies (■—■) of *Monoctonus pseudoplatani* at the woodland site in a year when there was no second peak of mummies (a) and when there was a second peak of mummies (b). The shaded block is the period during which young aphids should be abundant if there is to be a second peak.

abundance of the lime aphid on *Tilia cordata*. In addition, no parasitoid mummies were recorded during the 20-year population census of the Turkey oak aphid reported here (p. 93). Therefore, regulation by parasitoids is certainly not a general feature of aphid population dynamics.

Summarizing, there is no field evidence that unambiguously indicates that parasitoids are capable of regulating aphid abundance and there is good empirical support for the theoretical prediction that aphidophagous predators are ineffective at regulating aphid abundance.

INTRASPECIFIC COMPETITION

If individuals of a species compete with one another for a resource then this should affect their fitness in terms of contribution of individuals to future generations. Its occurrence is usually measured in terms of

Table 6.5. *The relationships between the logarithm of the number of immature aphids falling on to horizontal sticky traps (Y) and the logarithm of the number moving on to leaves (X) each week on two trees in 1968*

Instar	Tree	Relationship	n	r
First, second and third	A	log Y = 0.19 + 0.93 log X	24	0.94
	B	log Y = 0.33 + 0.88 log X	22	0.89
Fourth	A	log Y = 0.48 + 0.66 log X	24	0.63
	B	log Y = 0.69 + 0.43 log X	21	0.60

survival and/or fecundity. This implies that the resource for which they compete is limited. Finally the adverse effect of competition on the fitness of individuals is greater the greater the number of competitors. That is, the effect of intraspecific competition is density dependent. Here the evidence for intraspecific competition occurring in the field and its effect on sycamore aphid fitness is reviewed.

COMPETITION FOR SUITABLE SPACE

As indicated on pages 26–8 sycamore aphids of all stages frequently move from leaf to leaf and this results in the distribution between leaves continually changing both horizontally and vertically within the canopy (Dixon, 1969; Dixon & McKay, 1970). The rain of aphids from the canopy is closely associated with the incidence of movement of aphids between leaves. Aphids that fall off or are dislodged from leaves are at risk of death, especially if they are immature and unable to fly back up into the canopy. As the movement between leaves is risky it is relevant to ask: What is (are) the advantage(s) of moving? It is likely that it results in them occupying the most favourable positions on leaves or the most favourable leaves (p. 29). Interestingly, the number of immature stages falling or being dislodged from leaves each week, and becoming trapped on horizontal sticky traps placed beneath the trees, is either proportional or sub-proportional to the numbers moving from leaf to leaf (Table 6.5). That is, more movement between leaves does not result in a super-proportional response in terms of more falling off.

Leaves

Aphids space themselves out on leaves (p. 58). If preferred space is at a premium then one would expect aphids to accumulate in greater

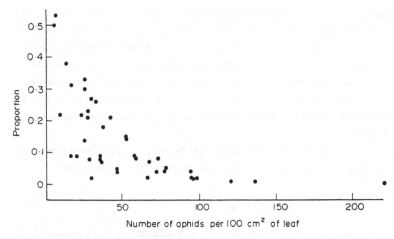

Figure 6.8. The relationship observed in the field between the
proportion of adult aphids at spacings greater than 1 cm and the
number of aphids per 100 cm² of leaf.

numbers per unit area in such areas. What evidence is there that they
compete for space?

The greater the number of aphids on a leaf at a particular time
the more closely they are packed together (Figure 6.8). That is, they
appear to have a preferred spacing but when abundant the aphids
pack more tightly onto the preferred leaf area. This indicates that with
increased density there is an increased competition for suitable space
(Dixon & Logan, 1972).

Tree

During summer, aphids aggregate on the leaves in the lower canopy
of sycamore trees. From August onwards there is a recolonization of
the upper canopy resulting in an equal distribution throughout the
canopy in early autumn in most years. Over the period 1966–1973 the
relationship between the logarithm of the number of aphids per 40
leaves in the upper canopy (Y) and the logarithm of the numbers
per 40 leaves in the lower canopy (X) in September is:

$$\log Y = -0.82 + 1.15 \log X$$
$$(r = 0.82, n = 8; b > 1, t = 0.46, \text{NS}) \tag{6.7}$$

Although the numbers per leaf in the upper canopy appear to be
increasing super-proportionally, the exponent 1.15, is not significantly

greater than 1. However, the ratio of the numbers per leaf in the upper and lower canopy in September in those years when the aphid was least abundant, and when it was most abundant, indicates a density dependent movement into the upper canopy (χ^2 based on totals = 453.6, and averages 113.2). In the four years when the aphid was least abundant, the numbers per leaf in the upper canopy was 0.3 of that in the lower canopy, and the four years when they were most abundant it was 0.8.

That is, both on leaves and trees there is good evidence that there is competition for suitable space when the aphid is abundant.

EFFECT ON BODY SIZE

Body size within a species in aphids is affected by temperature, food quality and crowding (Dixon, 1987c). Seasonal changes in temperature and food quality, therefore, tend to obscure the effects competition might have on body size. However, at a particular time of year sycamore aphids are strikingly smaller when abundant than when uncommon. For example, adult weight of first generation individuals, over a period of seven years, is negatively correlated with the logarithm of the peak number of fourth instar individuals (Figure 6.9). That is, after removing the confounding issues of food quality and temperature by comparing across years there is strong evidence for a density dependent effect of crowding on body size, with the aphids being much smaller at high than low population densities.

EFFECT ON REPRODUCTION

In the sycamore aphid, reproduction is proportional to their weight raised to the power of 1.5 if reproduction is measured in terms of biomass and 1.3 if measured in terms of number of offspring (Kindlmann et al., 1992). That is, a doubling in weight results in a 2.8 increase in the biomass and a 2.5 increase in the number of offspring produced.

The reproductive rate of sycamore aphids can be determined by keeping individual aphids in small leaf-cages. By using cages of the same area, half of which have a partition to keep the aphids apart, and placing two adult aphids in each cage it is possible to test whether the interaction between aphids results in a depression in the reproductive rate, or whether nutrition affects the interaction. This experiment was done when the leaves were young, mature and senescent (Table 6.6). In the absence of a partition in the cage, and when leaves are mature and

Table 6.6. *Reproductive performance of two aphids free to interact with one another expressed as a proportion of that of two aphids kept apart*

Date	Leaves	Relative reproductive performance	Significantly different from 1
23–31 May	Young	1.3	NS
6–18 July	Mature	0.26	$P < 0.001$
1–15 August	Senescing	1.1	NS

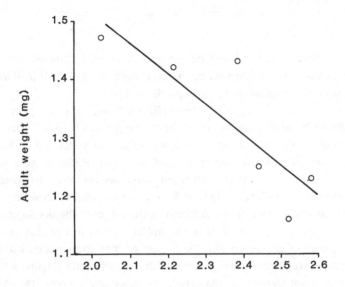

Figure 6.9. The relationship observed in the field between the average weight of first generation adults (Y) and the logarithm of the peak number of fourth instar aphids (X) that gave rise to the adults for one tree over a period of six years (Y = 2.56–0.52 log X).

nutrition poor, there is a pronounced reduction in the reproductive rate. This is not so when leaves are young or senescent. As the quantity of food and space per pair of aphids is the same in the two types of cage the reduction in the reproductive rate of the aphids in undivided cages on mature leaves must result from the interaction between the aphids. This was confirmed by using groups of aphids (two or six), while keeping the area of leaf available to each aphid constant. Increasing the size of the group from two to six has a highly significant effect in depressing reproductive rate (Table 6.7). This effect is still apparent, though

Table 6.7. *Reproductive performance of aphids kept in groups of six expressed as a proportion of that of aphids kept in groups of two. Both groups of aphids experienced the same density in terms of numbers of aphids per unit area of leaf*

Leaves	Relative reproductive performance
Mature	0.09
Senescing	0.38

less marked, after the leaves have begun to senesce. Thus interaction between aphids is influenced by the nutritive quality of their food and the number of aphids in the group (Dixon, 1970b).

Crowding also appears to be mediated through the host plant, as is shown by rearing aphids in or opposite a group of aphids. That is, a group is caged on the upper or lower surface of a leaf and an individual on the opposite side of the leaf. Both the relative growth and developmental rates are greatest when aphids are reared in or opposite small rather than large groups and the negative effect of crowding on the relative growth rate is significant. Although both the developmental and relative growth rates of the aphids reared in isolation opposite groups of aphids are consistently higher than those reared within groups, nevertheless, they do not differ significantly (Figure 6.10). Although not significant, the consistent difference supports the existence of a direct interaction between individuals and the similar effect obtained when reared in or opposite a group indicates an indirect interaction via the plant.

Recruitment

The reproductive rates of aphids on two trees growing on the suburban site were measured daily for eight years. Combined with the number of adult aphids on these trees the reproductive rates give an indication of the total recruitment of young aphids. In spring the recruitment is strongly inversely related to the number of young aphids colonizing the leaves at bud burst (Figure 6.11). Average daily reproduction is only one of the processes underlying this relationship. The migration (p. 90) and mortality (p. 89) of first generation adults also shapes this response. One

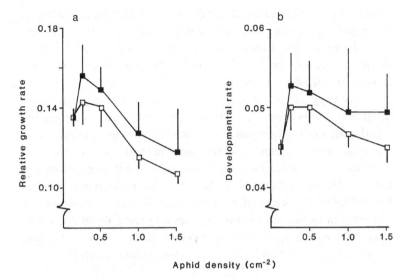

Figure 6.10. The relationships between (a) relative growth rate (RGR, μg^{-1} day^{-1}) $\pm$ s.e. and (b) developmental rate (1/days) $\pm$ s.e. and initial aphid density for aphids reared within crowds (□) and individuals reared opposite crowds, in isolation (■).

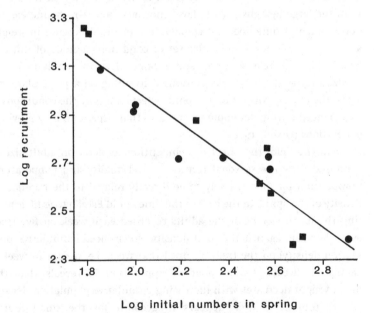

Figure 6.11. The relationship between recruitment in spring and the initial number of aphids colonizing the leaves at bud burst for the two sycamore trees (●, ■) on the suburban site over a period of eight years.

possible factor affecting the density dependent migration of adults is the negative effect the aphids colonizing the leaves at bud burst have on leaf size (p. 154). That is, not only are the aphids competing for resources they are also possibly reducing the carrying capacity of the leaves by reducing their size.

The peak number of aphids achieved in autumn depends on the number of adults present in early autumn and on their reproductive rate (Dixon, 1970b). The reproductive rate of aphids in early autumn as they come out of reproductive diapause varies from year to year. It is not correlated with the number of aphids present at that time or with temperature differences from year to year. However, it is high in years when few aphids are present earlier in the year and low when aphids are abundant in spring and summer (Figures 6.12 and 6.13; Dixon, 1975a). That is, there is evidence of a delayed density dependent effect of crowding in spring and summer on the reproductive rate in autumn.

CUMULATIVE POPULATION DENSITY

Food quality, temperature and competition for food all affect body size in aphids. Small nymphs born to small mothers take longer to mature than the large nymphs born to large mothers, however, the increase in developmental time does not compensate for the difference in weight at birth. Usually for a particular set of conditions a clone of initially large aphids decreases in size and a clone of small aphids increases in size over a number of generations converging on a particular size. Thus, the current weight of an aphid reflects not only the conditions it experienced during development but also that experienced by previous generations (Dixon, 1998).

In the field, the effect of competition for food on adult size is confounded by the seasonal trends in food quality and temperature. Competition for food is likely to be directly related to the population density of the aphid. In the case of the lime aphid *Eucallipterus tiliae* plotting the mean weight of the adults recorded each week on five trees over two years against the total density experienced (cumulative population density) on the trees up until that week, i.e. weight in week 7 against Σ (weeks 1–6 number of aphids cm^{-2}), reveals that the mean weight decreases with increasing cumulative population density (Figure 6.14). This even appears to account for the small fourth generation adults, which experience low population densities during their development. This result could be confounded by the seasonal trends in plant quality and temperature. In an attempt to correct for

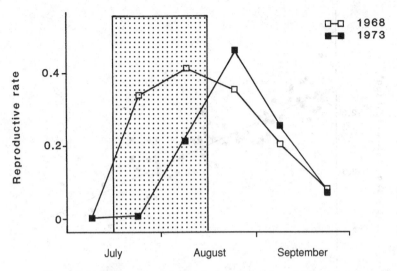

Figure 6.12. The number of offspring produced per adult per day on one tree in the months of July, August and September in 1968 and 1973. The stippled area is the period over which the autumnal reproductive period in Figure 6.13 was measured.

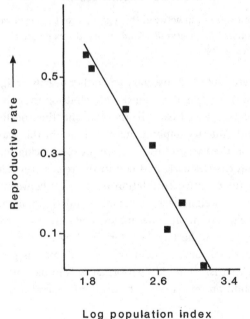

Figure 6.13. The relationship between autumnal reproductive rate (cf. Figure 6.12) and the logarithm of the average peak number of aphids present in spring and summer (April to 15 July) (population index).

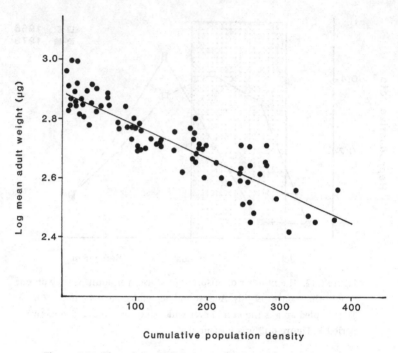

Figure 6.14. The relationship between the logarithm of the mean weight of adult lime aphids *Eucallipterus tiliae* and cumulative population density. (After Brown, 1975.)

this lime saplings were kept in an insectary and either heavily or lightly infested at bud burst. The population density of aphids on the initially heavily infested saplings peaked early in the year and then declined, whereas on the lightly infested saplings it peaked late in the year. In spite of the aphids on the two groups of saplings experiencing totally different population densities early and late in the year, nevertheless, there was a significant negative correlation ($r = -0.8$) between the logarithm mean of adult weight each week and cumulative population density. Thus the weight of adult lime aphids, and thus their reproductive potential, appears to be largely determined by the density of aphids present during their development and that of previous generations (Brown, 1975). That is, an aphid's performance is in part affected by a 'memory' of conditions experienced by previous generations.

EFFECT ON MORTALITY

Mortality during the development of nymphs reared within and opposite crowds increases as the density in the crowd increases (Figure 6.15).

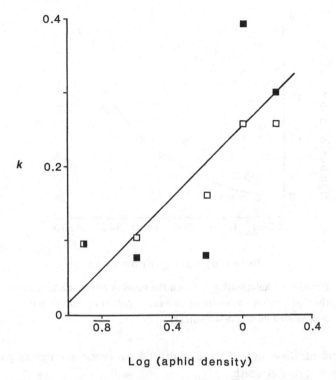

Figure 6.15. The relationship between the k-value for developmental mortality (log N_t/N_v) and initial crowd density (log N_t) for aphids reared within crowds (□) and opposite crowds (■), where N_t is the initial number of first instar aphids and N_v the number of aphids surviving to become adult.

Mortality rate is density dependent. That is, overall developmental mortality increases with developmental density, not because developmental rate tends to decrease with density (Figure 6.10b) but because the mean daily mortality rate increases significantly with density.

Adult mortality is more difficult to measure, especially in the field. Some indication of this, however, can be obtained from the deaths of adults that occur in the leaf cages used to monitor reproduction in the field. At the beginning of each week 10 adults were caged individually on each sample tree and the offspring they produced were removed and recorded each day for four days. This was repeated each week. The number of adults that died in the first three weeks in spring relative to the density they experienced during development is shown in Figure 6.16. In terms of percentage mortality the daily maximum is 40%. As with developmental mortality, adult mortality is also density

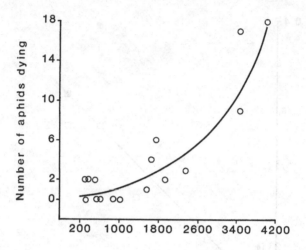

Figure 6.16. Relationship between the number of caged adults dying in the first three weeks of spring and the degree of crowding they experienced during development.

dependent. Dissection of these adults did not reveal any sign of parasitism or cause of death.

EFFECT ON MIGRATION

Studies on other aphids and on the sycamore aphid on sycamore saplings in the laboratory indicate that proportionally more aphids fly when numbers are high than when numbers are low (Dixon, 1969). Evidence for this occurring in the field is more difficult to obtain.

Spring migration

The relationship between the logarithm of the peak number of adult sycamore aphids in the lower canopy of sycamore and the logarithm of the peak number of fourth instar nymphs over the period 1960–68 has a slope of 0.41 (Figure 6.17). This indicates that proportionally more aphids migrate when aphid abundance is high than when it is low. Similarly, the same relationship for 10 trees in 1967 has a slope of 0.39, which again indicates density dependent migration. That is, there is good evidence for density dependent flight occurring in spring. Another maple aphid, *P. californiensis*, shows a similar density dependent spring

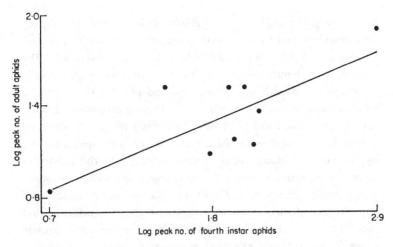

Figure 6.17. The relationship between the peak number of adults and the peak number of fourth instar aphids per 40 leaves, in spring, for one sycamore tree from 1960 to 1968.

migration and it is thought that this is the factor regulating its abundance (Furuta, 2003).

Summer migration

As for spring migration the slope of the relationship in summer is also significantly less than 1. For the period 1960–68 the slope is 0.44 and for 10 trees in 1967 it is 0.61. That is, there is similarly good evidence for density dependent flight in summer.

Autumn migration

Migration at this time of year is much more difficult to measure because generations three and four tend to overlap to a much greater extent than the spring and summer generations. Comparison of the average weekly suction trap catches (Y) with the average number of adults on the leaves each week (X) on two adjacent trees in the second week of September for the period 1965–73 gives the following relationship:

$$\log Y = -0.33 + 1.15 \log X \qquad (6.8)$$

The exponent 1.15, although larger is not significantly different from 1, therefore, there is no evidence for density dependent migration in autumn.

In summary, the sycamore aphid relative to its resource is generally very abundant and there are good theoretical grounds and empirical evidence that this is because its host, sycamore, is abundant. The variability in abundance of the aphid between trees appears mainly to be associated with tree phenology, and possibly the degree to which a tree is isolated from other trees. There is no evidence to indicate that the abundance of different but closely related tree-dwelling aphids is associated with their body size or per capita resource requirements. Although aphid-induced changes in the quality of sycamore in autumn were initially thought to regulate sycamore aphid abundance, several different experiments have failed to reveal any evidence of such an effect. Similarly, there is no evidence that natural enemies, although an important cause of mortality, regulate sycamore aphid abundance. They appear to be constrained by their dependence on the availability of young aphids, their relatively slow rate of development compared to that of the aphid, and the need to synchronize their life cycle with that of the aphid. The latter in particular results in the dynamics of the parasitoid often becoming uncoupled from that of the aphid in the second half of a year. Hyperparasitoids and predators also appear to inflict very heavy mortality on the parasitoids. There is, however, a lot of evidence to indicate that the sycamore aphid, when abundant, competes for suitable space and resources, and the effect of this competition is density dependent. It results in direct and delayed reductions in the rate of reproduction, and an increase in migration and mortality. In addition, these density dependent processes act continuously and often in parallel with one another.

7

Population dynamics

In spite of natural enemy-inflicted mortality initially being regarded as the most likely factor regulating tree aphid abundance this notion lacks empirical support. However, there is a lot of support for the idea that the abundance of tree-dwelling aphids is regulated by intraspecific competition (Chapter 6). Here an in-depth study of two population censuses is used to account for the different population patterns observed in two species of aphids, and more general population models presented, which incorporate the effect of cumulative density (Chapter 6) and seasonal changes in carrying capacity (Chapter 5), and both these factors. Finally, the data in the literature on other deciduous and coniferous tree-dwelling aphids is reviewed to determine the generality of regulation by intraspecific competition and the 'seesaw' effect – high numbers in spring followed by low numbers in autumn and vice versa.

PATTERNS IN POPULATION DYNAMICS

The studies on tree-dwelling aphids have revealed strong direct density-dependent recruitment and dispersal, and an inverse relationship between the numbers of aphids hatching from eggs in spring and the numbers present several generations later in autumn (Dixon, 1970c, 1971c,e; pp. 84–6). This delayed response is referred to as the 'seesaw' effect. The results of a more detailed analysis of two data sets are presented: the sycamore aphid *Drepanosiphum platanoidis* population census collected in Glasgow from 1960 to 1973 and that for the Turkey oak aphid *Myzocallis boerneri*, in Norwich from 1975 to 1995. The role of intraspecific competition operating via strong density-dependent migration and recruitment, and delayed density-dependent recruitment in regulating the abundance of these aphids is well documented. However,

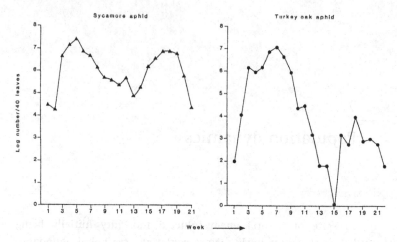

Figure 7.1. Within-year trends in the weekly numbers of aphids on a sycamore and a Turkey oak tree.

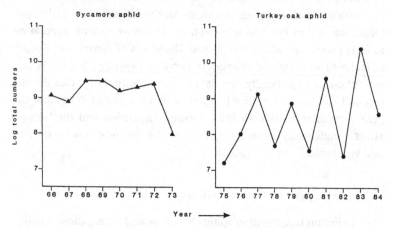

Figure 7.2. Between-year trends in the total numbers of aphids on a sycamore and a Turkey oak tree in Glasgow and Norwich, respectively.

these two species show very different within-year dynamics, with two similarly sized peaks in abundance in the sycamore aphid, one in late spring/early summer and the other in autumn; and predominantly only one peak in abundance in the Turkey oak aphid, which occurs in late spring/early summer (Figure 7.1). In addition, the between-year dynamics of the two species differ greatly, with relatively constant yearly totals in the sycamore aphid, and a more marked cyclical pattern in the yearly totals in the Turkey oak aphid (Figure 7.2). Interestingly, a similar but

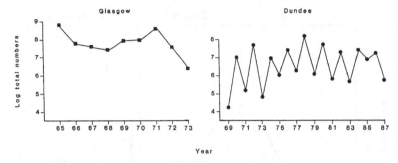

Figure 7.3. Between-year trends in the total numbers of sycamore aphids caught each year in a suction trap positioned 1.4 m above the ground close to two of the trees sampled in Glasgow, and in the Rothamsted Insect Survey trap positioned 12.2 m above ground and located at Dundee.

more chaotic cyclical pattern occurs in the annual catches of yellow water traps of the oak aphid *Tuberculoides annulatus*, collected over a period of 13 years (Robert & Rouzé-Jouan, 1976).

The relative rarity of sycamore aphids in 1973 (Figure 7.2) resulted from the still autumn of 1972 when the aphid achieved a very high level of abundance and was, as a consequence, abundant the following spring. As in 1961, the aphid declined dramatically in abundance in 1973 following the very high population in spring, giving a low overall abundance in 1973 (Dixon, 1979). This occurred in the almost complete absence of insect natural enemies. The otherwise relative stability in sycamore numbers on the trees from year to year is strikingly different from the regular cyclical pattern observed in the suction trap catches, which measure the aerial population of this species (Dixon, 1990b; Figure 7.3). This cyclical pattern proved irresistible to theoreticians (Turchin, 1990; Turchin & Taylor, 1992).

The objective here is to explain why the patterns in the population dynamics of these two species differ, and account for the apparent discrepancy in the between-year dynamics of sycamore aphid numbers in the air and on sycamore.

Within-year dynamics

The within-year dynamics of both species are largely determined by seasonal changes in host quality, with the leaves of trees most suitable for aphids in spring and autumn (pp. 24–5). This seasonality confounds population analyses, which may be complicated further by the length

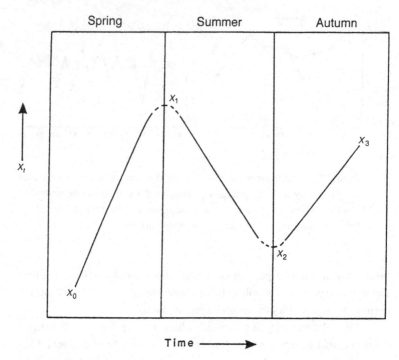

Figure 7.4. Diagrammatic representation of the within-year population dynamics.

of the vegetative season varying from year to year. To overcome this problem and to address the differences in the population profiles shown by the two species, each year is divided into a spring increase, a summer decline and an autumn increase (Figure 7.4). The latter is much weaker in the Turkey oak aphid than in the sycamore aphid (Figure 7.1). The rate of increase or of decline in each of these periods was determined by regression.

Definitions

Spring rates of increase of the adult population each year are the slopes of the relationships between the natural logarithms of the numbers of adults and time. The first point is that when adults are first recorded in a year. The last point is that which gives the longest period during which the growth is not significantly different from exponential growth for calculation of the growth rate. This method was adopted because, as the population approaches the summer peak (x_1), the population growth becomes sigmoidal rather than exponential. The rates of the summer declines are the slopes of the relationships between the natural logarithm of the numbers of adults and time. The first value is the first measurement following the summer peak (x_1) and the length of the series is again derived as above. This is necessary because population change in the summer trough in abundance is also sigmoidal rather than negative exponential (Dixon & Kindlmann, 1998b). The rates of increase in autumn are calculated in the same way as the spring rates of increase. The shape of the relationship between the numbers in autumn and those present the following spring are also determined by regression.

Rates of increase

The spring rates of increase are always negatively correlated with initial numbers of adults, and the summer rates of decline are also negatively correlated with the numbers of adults at the instant when summer migration starts, i.e. when the population curve becomes sigmoidal rather than exponential. In addition, the autumnal rates of increase are negatively correlated with the minimum number of adults in summer. These results are independent of species and site (Table 7.1, Figure 7.5).

A model of the within-year dynamics using the above relationships is derived as follows: if $x(t)$ is the natural logarithm of the population density at time t; x_0 is the natural logarithm of the population density in spring; s_1, s_2 and s_3 are the rates of increase before the summer peak is reached, of the subsequent decline, and of the autumnal increase, respectively; x_1 is the natural logarithm of the population density when the summer migration begins, which here for simplicity coincides with the summer peak population density; and x_2 is the population density at the end of summer, then

$x(t) = x_0 + s_1 t$ before the peak is reached,

$x(t) = x_1 + s_2 t$ during summer, and

$x(t) = x_2 + s_3 t$ in autumn

where t denotes time from the beginning of the corresponding period. Because of the inverse relationships between the spring rates of increase, summer rates of decline and autumn rates of increase, and initial numbers, $s_1 = k_1 x_0 + q_1$, $s_2 = k_2 x_1 + q_2$ and $s_3 = k_3 x_2 + q_3$, where k_1, k_2 and k_3 are negative slopes, and q_1, q_2 and q_3 the intercepts of the

Table 7.1. Slopes (k_i), intercepts (q_i), correlation coefficients (r_i) and their significance levels (P) of the relationships between natural logarithm of numbers of adults at the beginning of spring, summer and autumn, and the subsequent rate of increase or decline

	Site	Spring increase				Summer decline				Autumn increase			
		k_1	q_1	r_1	P	k_2	q_2	r_2	P	k_3	q_3	r_3	P
Sycamore aphid	1	-0.50	3.85	-0.92	<0.001	-0.17	-0.05	-0.067	<0.05				
	2	-0.43	3.14	-0.96	<0.001	-0.29	1.56	-0.70	<0.05	-0.34	2.68	-0.97	<0.01
Turkey oak aphid	1	-0.10	0.95	-0.61	<0.01	-0.15	0.02	-0.45	<0.05	-0.19	0.8	-0.17	NS
	2	-0.11	0.90	-0.64	<0.01	-0.10	-0.32	-0.42	<0.05	-0.01	0.71	-0.03	NS

NS, not significant

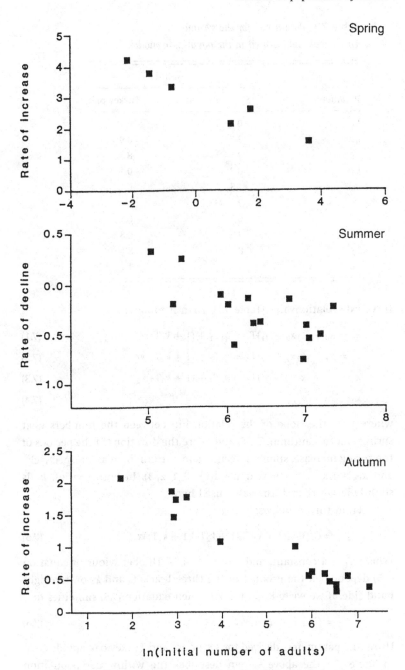

Figure 7.5. The rates of increase and decline achieved on two trees relative to the initial numbers of sycamore aphid adults present at the beginning of spring, summer and autumn, respectively.

Table 7.2. *Parameters for the sycamore and Turkey oak aphids used in the population model*

	Aphid	
Parameters	Sycamore	Turkey oak
k_1	−0.4	−0.11
q_1	3.5	0.9
T_1	2	8
k_2	−0.3	−0.1
q_2	1.5	0
T_2	2	8
k_3	−0.3	−0.1
q_3	2.7	0.8
T_3	8	8
k_4	1	1

three rate relationships (Table 7.1). Then it follows:

$$x_1 = x_0 + \{k_1 x_0 + q_1\} T_1 = q_1 T_1 + (1 + k_1 T_1)x_0 \tag{7.1}$$

$$x_2 = x_1 + \{k_2 x_1 + q_2\} T_2 = q_2 T_2 + (1 + k_2 T_2)x_1 \tag{7.2}$$

$$x_3 = x_2 + \{k_3 x_2 + q_3\} T_3 = q_3 T_3 + (1 + k_3 T_3)x_2 \tag{7.3}$$

$$x_{0,n+1} = k_4 x_{3,n} \tag{7.4}$$

where k_4 is the slope of the relationship between the numbers next spring and this autumn; T_1, T_2 and T_3 are the durations of the periods of the spring increase, summer decline and autumn increase, respectively; and $x_{i,n+1}$ means x_i in year $n + 1(i = 0, 1, 2, 3)$. In both species, k_4 is slightly larger than 1, the value used here.

From this it follows:

$$x_{i,n+1} = C_i + k_4(1 + k_3 T_3)(1 + k_2 T_2)(1 + k_1 T_1)x_{i,n} \tag{7.5}$$

Where C_i is a constant, and $i = 0, 1, 2, 3$. The behaviour of equation (7.5) depends on the product of the three brackets and k_4 on the right hand side. If we write $B_i = 1 + k_i T_i$, then equation (7.5) simplifies to

$$x_{i,n+1} = C_i + k_4 B_3 B_2 B_1 x_{i,n} \tag{7.6}$$

Using the parameters derived for sycamore and Turkey oak aphids cited in Table 7.2, the above system describes the within-year population trends observed in the two species (Figure 7.6). High numbers at the beginning of spring, summer and autumn are associated with low rates of increase and high rates of decline, and vice versa. The net effect of

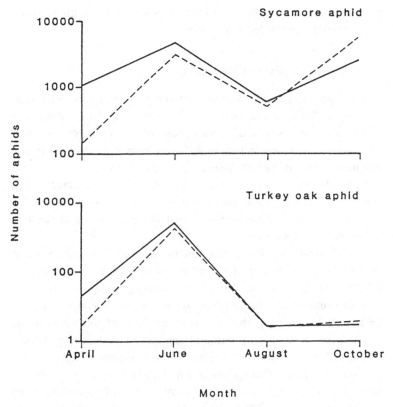

Figure 7.6. The trends predicted by equations (7.1), (7.2) and (7.3) in the numbers of sycamore and Turkey oak aphids in years starting with high numbers (solid line) and low numbers (dashed line).

this is that the sycamore aphid shows a strong 'seesaw' in abundance between spring and autumn.

In addition, the above also helps to resolve a paradox. The density-dependent effects of intraspecific competition on the population rate of increase are all undercompensating yet it is clear that in terms of the within-year dynamics the overall density-dependent effect is either compensating or overcompensating. The different density-dependent effects were measured over only a short period of each season. However, they act sequentially and/or in parallel continuously over the whole season. This notion is incorporated in equations 7.1–7.3 in the form of the multipliers T_1, T_2 and T_3, the duration of the spring increase, summer decline and autumn increase, respectively. The net effect of these multipliers, in terms of within-year dynamics, is to convert some or all of the undercompensating k_1, k_2 and k_3 factors into compensating or over-compensating density-dependent factors.

One or two peaks in abundance?

In both species of aphid there is a tendency for numbers to increase in autumn. The autumnal peak in abundance in the Turkey oak aphid is always at least one and often two orders of magnitude smaller than the summer peak, which is very different from that observed in the sycamore aphid where both peaks are similar in size (Figure 7.1). This could be due to the nutritional quality of the host, reflected in the reproductive rate of the aphid in autumn, differing on sycamore and Turkey oak, and/or the extent to which the populations of the two species decline in summer.

The trends in soluble nitrogen in the leaves of sycamore and Turkey oak (an indicator of nutritive quality) are similar in that it is high in spring when the leaves are actively growing, falls to a low level in summer, and increases in autumn prior to leaf fall (Dixon, 1963, 1971c). Accepting that the concentration of soluble nitrogen in the leaves is a good indicator of changes in host quality, Turkey oak should be a better host for aphids than sycamore. However, on sycamore there is another species of *Drepanosiphum* – *D. acerinum* – which continues to reproduce all through summer when *D. platanoidis* is in reproductive diapause. Thus, although soluble nitrogen levels give an indication of changes in host quality in time, they cannot be used as an absolute indicator of host quality for a particular species of aphid (p. 25).

A more direct measure of host quality is given by the reproductive rate achieved by the aphid. This has been measured for both species by clip-caging adult aphids on the leaves of their host trees throughout the course of several years in Glasgow. In addition, the Glasgow census data on the two species include a record of the numbers of small nymphs and adults present each week. The ratio of the number of small nymphs in a week to the number of adults in the preceding week is correlated with the number of offspring produced per adult per day over the same period, for both sycamore and Turkey oak aphids ($r = 0.79$, $P < 0.001$; $r = 0.89$, $P < 0.001$, respectively). Thus the ratio of small nymphs to adults can be used as a measure of the reproductive rates achieved, and allows comparison of the seasonal trends in reproductive performance of the sycamore aphid in Glasgow and the Turkey oak aphid in Norwich. This indicates that the Turkey oak aphid has a higher reproductive rate

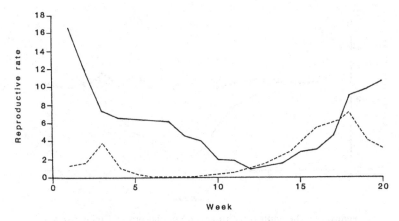

Figure 7.7. The seasonal changes in the reproductive rates achieved by sycamore aphids (dashed line) and Turkey oak aphids (solid line).

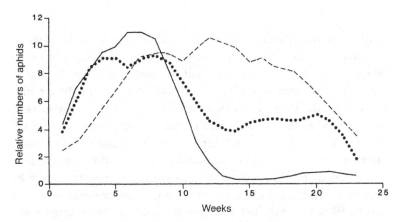

Figure 7.8. The within-year population profiles of Turkey oak aphids on single trees in Glasgow (dotted line) and in Norwich (solid line), and on caged saplings (dashed line), each scaled so that the peak numbers are similar.

than the sycamore aphid in spring and early summer, but a similar rate in autumn (Figure 7.7). Therefore, the small autumnal peak in the abundance of the Turkey oak aphid, relative to that achieved by the sycamore aphid, cannot be attributed to the latter having a higher rate of increase in autumn.

Interestingly, the population trends of the Turkey oak aphid on the two trees in Norwich, on the one tree in Glasgow and on the caged saplings differ (Figure 7.8). This cannot be accounted for in terms

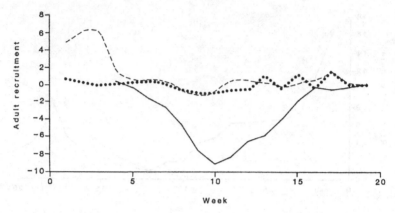

Figure 7.9. The adult recruitment to populations of Turkey oak aphids on single trees in Glasgow (dotted line) and in Norwich (solid line), and on caged saplings (dashed line).

of nymphal recruitment rates as the seasonal trends and rates were very similar in all three cases. However, plotting the changes in numbers of adults between one week and the next divided by the number of large nymphs in the first week indicates that, in the middle of the year, the trees in Norwich lost more adults than they gained by recruitment. This is in marked contrast to what was observed in Glasgow and on the caged saplings (Figure 7.9). Caging clearly reduces the losses due to natural enemy activity and migration. The natural enemy activity on the Glasgow tree was greater than on the Norwich trees but, possibly more importantly, the Glasgow tree was enclosed on two sides by a very tall building, the walls of which often took on a yellow coloration from the countless numbers of Turkey oak aphids that settled there during the migratory period in summer. It is likely that many of these aphids subsequently flew back on to the tree. That is, it is likely that the presence of high walls on two sides of the tree severely restricted migration. The small autumnal peak in the Turkey oak aphid in Norwich is mainly determined, therefore, by the very low level to which the population declines in summer (Figure 7.1). The extent to which populations of this species decline in summer appears to depend on the extent to which the host tree is isolated from other trees. As sycamore is far more abundant than Turkey oak, sycamore aphid populations do not decline to such low levels in summer, and consequently can achieve high levels of abundance in autumn.

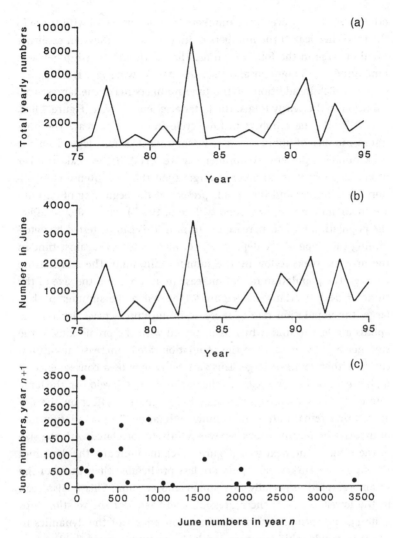

Figure 7.10. The total number of Turkey oak aphids recorded yearly and in June, and the relationship between numbers of aphids in June for that year and the following year, 1975–95.

BETWEEN-YEAR DYNAMICS

On trees

The total numbers of aphids counted on sycamore from year to year vary less than those on Turkey oak (Figure 7.2), which tend to cycle in abundance. This cyclical pattern is seen even more strikingly in the year-to-year changes in the June numbers (Figure 7.10b), but less so in

other months. However, June numbers are not correlated with those in
the following year. If the numbers in June are small, they can be either
small or large in the following June. That is, the relationship between
June numbers in one year and those in the following year is triangular
(Figure 7.10c). In addition, as the June numbers make the largest contribution to the yearly totals, the latter also tend to cycle (Figure 7.10a).

Why do the numbers in June cycle from year to year? The minimum number of aphids present in summer does not depend on the
peak numbers present in summer (Figure 7.11a). If the peak number
of aphids present in summer is large, then there is intense competition for resources and the aphids present at the beginning of autumn
are small and have a low fecundity (Dixon, 1990b), which should affect
the population rate of increase in autumn. This can be tested by determining the slope of the dependence of ln(density + 1) against time in
the first six weeks following the summer minimum. The relationship
between the population rate of increase in autumn and the size of the
summer peak is triangular (Figure 7.11b). When the summer peak is
large, the population rate of increase in autumn is never large – the
aphids are less fecund, which is in accord with the predictions. If the
summer peak is small, then the population rate of increase in autumn
can be either small or large. This variability may be a consequence of
high predation in some years, as the predators that develop on the large
numbers of aphids in summer may have a marked effect on the few
aphids that remain after the summer migration. The population rate
of increase in autumn, however, was positively correlated with the size
of the peak in the next year (Figure 7.11c). In summary, the dynamics
shown by the Turkey oak aphid are less predictable than those of the
sycamore aphid, especially in summer and autumn. This is also seen
in the lower values of the regression coefficients of the relationships
defining the spring, summer and autumn phases of the dynamics in
the Turkey oak aphid compared with the sycamore aphid (Table 7.1).

In order to take this stochastic behaviour into account, the
between-years population dynamics of the Turkey oak aphid can be
described by equations (7.1) to (7.6), but with equations (7.2) and (7.3)
replaced by:

$$x_2 = x_1 + (k_2 x_1 + q_2)T_2 + r.\text{RND}_1 = q_2 T_2 + (1 + k_2 T_2)x_1 + r.\text{RND}_1$$

$$(7.2a)$$

$$x_3 = x_2 + \text{RND}_2(k_3 x_2 + q_3)T_3 = \text{RND}_2 q_3 T_3 + (1 + \text{RND}_2 k_3 T_3)x_2$$

$$(7.3a)$$

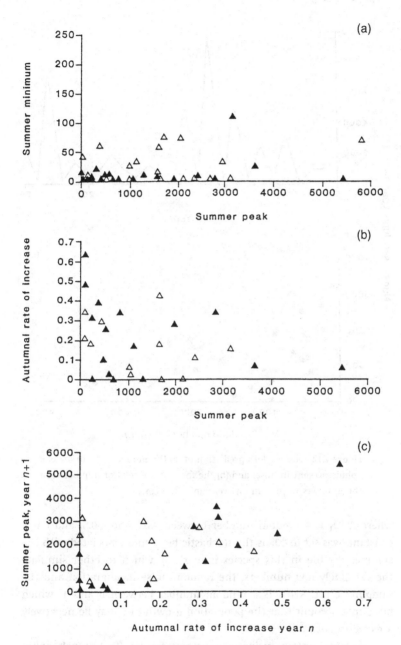

Figure 7.11. The relationship between (a) the minimum number of
Turkey oak aphids in summer and the summer peak numbers; (b) the
autumnal rate of increase and the summer peak numbers; and (c) the
summer peak numbers for one year and the autumnal rate of increase
the previous year, for two trees (closed and open triangles, respectively).

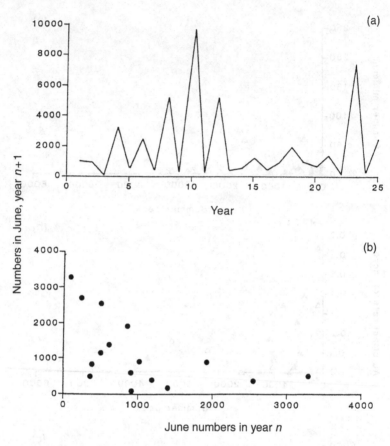

Figure 7.12. The model's prediction of (a) the numbers of Turkey oak aphids present in June; and (b) the form of the relationship between the numbers in June in one year and the next.

where RND_1 is a random number between -0.5 and $+0.5$, and r is a constant (was set to 10, as the stochastic behaviour associated with the summer decline in this species is marked), which together simulate the extremely low numbers, the random noise in migration and the sampling error. RND_2 is a random number between 0 and 1, which takes into account that the population growth rate may be negatively affected in autumn.

In this system, using the parameters for Turkey oak aphid (Table 7.2), the numbers in June also cycle (Figure 7.12a) and the plot of numbers in June against numbers the preceding June is triangular (Figure 7.12b). That is, the cyclical pattern is driven by the inverse relationship between the size of the spring peak and the population rate of increase in autumn. Interestingly, in the absence of noise in the

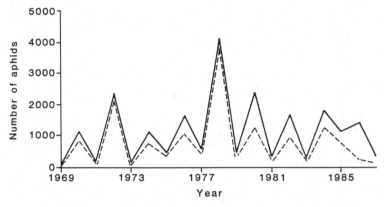

Figure 7.13. The total numbers (solid line) of sycamore aphids caught each year from 1969 to 1987, and the numbers caught in June, October and November each year (dashed line) over the same period by the suction trap located at Dundee.

system, the cyclical pattern is more definite and the amplitude of the fluctuations smaller. In addition, running the system for periods equivalent to 200 years reveals a weak but significant 'seesaw' effect.

The trend in sycamore aphid numbers can be predicted from the above equations, using the parameters for the sycamore aphid (Table 7.2), and making $r = 1$. The latter can be justified as, although there is noise in the sycamore aphid system, mainly a consequence of variability between years in the wind speed in autumn (p. 40; Dixon, 1979), the noise is considerably less than in the Turkey oak aphid system. The prediction of the model is that there should be little change in sycamore aphid numbers from year to year (see Figure 7.15). This stability is attributable to the two within-year peaks in abundance being similar in size and both making a major contribution to the yearly totals. The inverse relationship between the size of the spring and autumn peaks is such that a change in the size of one peak is compensated for by a similar but opposite change in the other.

In the air

The yearly catches of the suction trap positioned close to two of the sycamore trees closely reflected the total average abundance of the aphids on those trees, changing relatively little from year to year (Figure 7.2). In contrast, the yearly catches of the Rothamsted Insect Survey trap at Dundee fluctuated markedly from year to year (Figure 7.3). The aphids caught in June, October and November mainly determine the size of the annual catch of the Dundee trap (Figure 7.13). The aphids

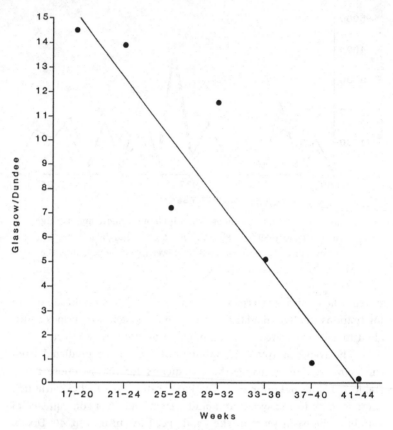

Figure 7.14. The relationship between the ratio of the numbers of sycamore aphids caught in the Glasgow and Dundee suction traps and time in weeks from the beginning of a year.

caught in June make up only 25% of the total caught in these three months. The sum of the catches in these three months is well correlated with the yearly totals ($r = 0.95$, $P < 0.01$), which also show similar dynamics, i.e. they cycle (Figure 7.13). Therefore the Dundee trap mainly catches aphids late in a year. In addition, the ratio of the catches of the local and the Dundee trap each month do not remain constant. Early in a year, the local trap catches many more aphids relative to the Dundee trap than late in a year (Figure 7.14). This change in flight behaviour can be represented by a correction factor, c, which increases in value from 0.08 to 1 from spring to autumn. Biologically this makes sense as early in a season not all trees are equally infested with aphids, mainly because of differences in tree phenology and size (p. 124). At this time

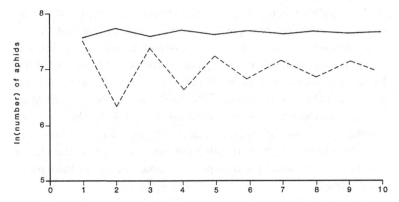

Figure 7.15. The model's predictions of the year-to-year changes in the total numbers of sycamore aphids on trees (solid line) and caught in the Dundee suction trap (dashed line).

it is possibly more advantageous for aphids to redistribute themselves locally between trees and colonize the smaller trees. Later in the year there are advantages in the aphids flying higher and colonizing the taller trees (p. 134). As a consequence these aphids are more likely to be carried by prevailing winds and caught by the Rothamsted Insect Survey traps, which are positioned 12.2 m above the ground and well away from woodland.

Accepting that the dynamics of the sycamore aphid can be represented by equations (7.1) to (7.6), with the modification in equations (7.2a) and (7.3a), it is possible to predict the relative trends in the yearly sizes of total tree counts and Dundee suction trap catches. The tree count will be proportional to $x_0 + x_1 + x_2 + x_3$, and the trap catch to $c_0 x_0 + c_1 x_1 + c_2 x_2 + x_3$, where c_0, c_1 and c_2 are $\ll 1$. Using these relationships, the prediction is that the trap catch will oscillate from year to year, but the tree count will not (Figure 7.15). This conforms with what is observed (see Figs. 7.2 and 7.3).

THEORY

The above models accurately mimic the within- and between-year trends in the numbers of individuals of two species of aphid. The task now is to move from the specific to the general. The optimum model for this is one that is simple and includes little.

The stability of animal populations is accounted for in one of two ways. Regulation is relative to an equilibrium density and the direction of population change at any one time is determined largely by the size of the population relative to the equilibrium (Nicholson, 1933), that is, regulation is density dependent (Smith, 1935). Alternatively, population change is usually random with respect to population size, but constrained by an upper limit imposed by the carrying capacity of the habitat and the lower limit of extinction (Andrewartha & Birch, 1954; Milne, 1957a,b). That is, trends in abundance are either determined by density-dependent feedback mechanisms, or mainly by the organism coming up against the carrying capacity of their habitat. These concepts have frequently been revisited but not reconciled. In the case of aphids, however, both appear to play an important role in their population dynamics.

CUMULATIVE DENSITY MODEL

In this model the regulatory term that slows down the instantaneous rate of increase is cumulative density (p. 86), rather than a term that is proportional to instantaneous density, as in logistic growth. This is based on the assumption that it is the sum of the numbers of individuals times their life spans that determines the slowing down of the instantaneous rate of increase. The logistic growth never yields a decline in population density in time – a phenomenon typical of aphid population dynamics. In the absence of an effect of natural enemies or change in carrying capacity an autoregulatory term that causes a decline in population density in time is needed. Cumulative density is a potential candidate, as there is good empirical evidence that it can slow down population rate of increase (p. 86).

Accepting this, aphid population dynamics can be described by the following set of differential equations:

$$\frac{dh}{dt} = ax, \qquad h(0) = 0, \qquad (7.7a)$$

$$\frac{dx}{dt} = (r - h)x, \quad x(0) = x_0, \qquad (7.7b)$$

where equation (7.7a) describes the temporal changes in cumulative density, $h(t)$, of aphids at time t, and equation (7.7b) describes the temporal changes in aphid density, $x(t)$, at time t, a is a scaling constant relating aphid cumulative density to its own dynamics, and r is the maximum potential growth rate of the aphids. Thus, while, for example in the logistic model it is assumed that the growth rate linearly declines with population density, which is described by the term $(1 - r/K)$ in the logistic equation, here we assume that the growth rate linearly declines with aphid cumulative density, which is expressed by the term $(r - h)$. This models the decline in population density with time, so typical of aphid populations.

Model analysis

The system of differential equations (7.7a,b) can be analytically treated as follows: dividing (7.7b) by (7.7a) gives,

$$\frac{dx}{dh} = \frac{r - h}{a},$$ (7.8)

and separation of variables in (7.8) yields an explicit solution – aphid density as a function of its cumulative density:

$$x(h) = \frac{h(2r - h)}{2a} + x_0$$ (7.9)

Equalizing (7.9) to 0, $x(h) = 0$, gives,

$$h_1 = r - \sqrt{r^2 + 2ax_0}, \; h_2 = r + \sqrt{r^2 + 2ax_0}$$ (7.10)

Expression (7.9) is a parabola, which in the $x - h$ plot crosses the vertical x-axis at x_0 and the positive part of the horizontal h-axis at the point h_2 defined by (7.10).

Setting (7.8) equal to 0, $\frac{dx}{dh} = 0$, gives the h-coordinate of the peak of this parabola:

$$h_{max} = r.$$ (7.11)

The vertical coordinate of the peak of this parabola can be found by substitution of (7.11) into (7.9):

$$x_{max} = \frac{r^2}{2a} + x_0$$ (7.12)

Thus the aphids increase in numbers by $\frac{r^2}{2a}$ from an initial value at the beginning of the season. The height of the peak is positively correlated with the initial density of aphids, their intrinsic rate of increase and negatively with the cumulative density parameter a. The peak is reached when the cumulative density is equal to the aphid intrinsic rate of increase. The former, however, is positively correlated with the initial aphid density. Therefore, with increasing initial aphid density, the peak density becomes larger, and is achieved earlier. This is a characteristic feature of the spring peak in abundance of tree-dwelling aphids and is illustrated in Figure 7.16. That is, this model successfully simulates one

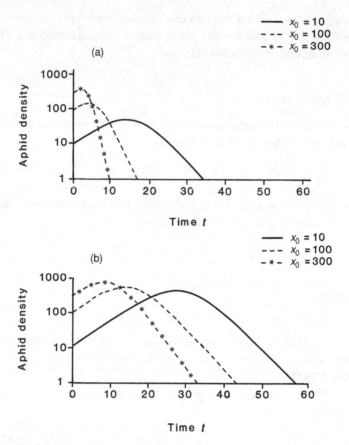

Figure 7.16. Prediction of the trends in aphid population density as a function of time for various initial population densities of the aphid. (Equations (7.7a,b): $r = 0.3$; (a) $a = 0.0005$; (b), $a = 0.00005$; initial aphid densities (x_0) in key).

of the important features of aphid dynamics. One example of the fit of this model to empirical data is shown in Figure 7.17.

LOGISTIC MODEL WITH VARIABLE CARRYING CAPACITY AND GROWTH RATE AFFECTED BY CUMULATIVE DENSITY

Instead of assuming as in the logistic model that the carrying capacity remains constant it is more realistic to vary the carrying capacity in time. In the case of sycamore the carrying capacity appears to be high early and late in a year and lowest in late summer (Figure 5.6). In addition, empirical data indicates that if aphids are abundant early

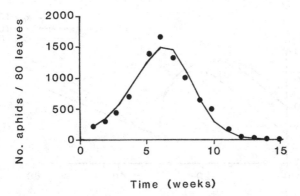

Time (weeks)

Figure 7.17. Aphid density as a function of time predicted by equation (7.7) (solid line), relative to the census data for the aphid on a Turkey oak tree in 1995 (•); estimated parameters: $r = 0.71$; $a = 0.00015$, initial aphid density $x_0 = 215$.

in a season there is intense competition for resources and the aphids present at the beginning of autumn are small and have a low fecundity, which affects the population rate of increase in autumn (Dixon, 1975a, 1990b). That is, population growth rate is negatively affected by cumulative density (p. 86). These two concepts can be incorporated into the model as follows:

$$\frac{dh}{dt} = ax, \qquad\qquad h(0) = 0, \qquad\qquad (7.13a)$$

$$\frac{dx}{dt} = (r - h)x\left(1 - \frac{x}{K}\right), \quad x(0) = x_0, \qquad\qquad (7.13b)$$

where $K = K(t) = (K_{max} - K_{min}) \cdot ((\cos(t^d) + 1)/2) + K_{min}$.

Equation (7.13a) is the same as equation (7.7a) in the previous model and describes changes in aphid cumulative density, but in contrast to the previous model, equation (7.13b) is a combination of equation (7.7b) and the classical logistic equation, which means that both cumulative density (via term $(r - h)$) and instantaneous density (via term $(1 - x/K)$) have a negative effect on aphid growth rate. This model, unlike the logistic model, incorporates the variable carrying capacity (K) described on page 62, and yields a 'seesaw' effect (Figure 7.18). It is also possible that the temporal trend in aphid abundance also affects the trend in carrying capacity. However, the general outcome is the same.

The above model possibly resolves another puzzling feature of sycamore aphid population dynamics. The inverse within-year relationship between autumn and spring abundance clearly differs between trees in terms of the intercepts. The above model predicts this if the

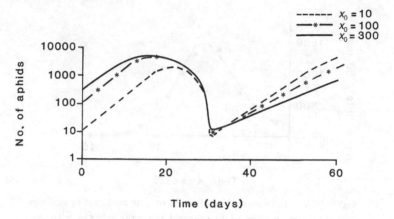

Figure 7.18. Predicted trends in aphid population density as a function of time for various initial densities of the aphid. Prediction of the logistic model with variable carrying capacity: $r = 0.3$; $K_{max} = 10\,000$, $K_{min} = 1$; $d = 33$ days; initial aphid densities (x_0) in key.

maximum potential rate of increase (r) is varied in equation 7.13b. It is difficult to correct for the confounding effect of population density on reproductive rate. However, the little empirical evidence there is tends to support the idea that r varies between trees.

APHID POPULATION DYNAMICS AND THE 'SEESAW' EFFECT

The above studies on the sycamore and Turkey oak aphids, and an auto-regression analysis of the Turkey oak aphid census data (Sequeira & Dixon, 1997), indicate that regulation of their abundance occurs within rather than between years. By dividing each year into three phases, spring increase, summer decline and autumn increase (Figure 7.4), it is possible to avoid the confounding effect of seasonality. The population increases, decreases and increases in these three phases, respectively, irrespective of aphid abundance. That is, there is an equilibrium trajectory rather than an equilibrium point. What is important for understanding the population dynamics is how the rates of increase and decline are related to aphid abundance.

Intraspecific competition in each of these phases results in direct density-dependent effects on adult size, recruitment and migration (pp. 82, 84–6, 90), which operate via rates of increase and decline to regulate aphid abundance. In addition, there is a delayed density-dependent response that operates between phases, with high peaks of abundance in spring/summer reducing the rate of increase in autumn (pp. 84–86).

A marked feature of the population dynamics of deciduous tree-dwelling aphids is the 'seesaw' effect. This is the negative correlation between numbers of aphids in the first generation in spring and the last generation in autumn, which is present in lime aphid (Dixon, 1971e), pecan aphid *Monellia caryella* (Liao & Harris, 1985), maple aphid *Periphyllus testudinaceus* (Shearer, 1976), sycamore aphid and Turkey oak aphid. However, the effect varies in strength between species; in the two species studied in detail, it is strong in the sycamore aphid and weak in the Turkey oak aphid. In a 13-year study of the population dynamics of another maple aphid, *P. californiensis*, the three years with the highest spring peak numbers had low autumn numbers and high autumn numbers always occurred in years with low spring numbers. Overall, however, the numbers were not significantly correlated (Furuta, 2003). This aphid, like the Turkey oak aphid, has predominantly only one peak in abundance each year, which for the reasons given on page 104 might have obscured the 'seesaw' effect in this case.

In those species in which the 'seesaw' effect is strong, it appears to be driven mainly by one component in the yearly cycle – the autumnal rate of increase. This could operate in one of several ways. High aphid numbers early in a year adversely affect the quality of: (a) the host plant in autumn; (b) the aphid; or (c) both the aphid and the plant. High numbers of the pecan aphid early in a year adversely affect the quality of pecan for the aphid later in the year (Wood *et al.*, 1985; Alverson & English, 1990; Bumroongsook & Harris, 1991). In the sycamore aphid the effect appears to operate through the aphid, with high aphid abundance early in a year resulting in the production of very small aphids, which are late in coming into reproduction in autumn and have a low reproductive rate (Dixon, 1975a; Figure 6.12; Chapter 6). There is no evidence that a high abundance of this aphid in spring induces changes in sycamore that are detrimental to the aphid in autumn (pp. 70–1). The Turkey oak aphid shows a weak 'seesaw' response. That is, a general feature of the deciduous tree aphid system is a within-year 'memory' that transfers information on abundance from spring to autumn.

Unexpectedly, the differences in the shapes of the yearly population profiles of the two species studied in detail here cannot be attributed to differences in the nutritional quality of the host trees. In the Turkey oak aphid, the marked and consistent asymmetry in the sizes of the summer and autumnal peaks appears to be determined by the large migratory losses during summer. It is likely that, in a Turkey oak forest, trees would gain as well as lose aphids, and the net loss in summer is likely to be less than from isolated trees, like those in Norwich. In other aphid–tree systems, an abundance of aphids early in a year renders the host less suitable for aphids later in the year. This appears to be the case for the pecan aphids (Liao & Harris, 1985; Alverson & English, 1990). The yellow pecan aphid *Monelliopsis pecanis* and the black-margined aphid *Monellia caryella* feed on different parts of pecan leaves (Tedders, 1978). Nevertheless, prior exposure of leaves of pecan to high densities of aphids of either species has an asymmetrical impact on the population growth of the other species (Bumroongsook & Harris, 1992). That is, through their effect on the host these aphids can adversely affect the population growth of one another. In other species, such as the lime aphid, there is evidence that the host is similarly affected (Barlow & Dixon, 1980), but much less so than the pecan. In systems where aphids have a marked adverse effect on the quality of their host plant, a single peak in abundance is to be expected in those years when aphids are abundant in spring. Furthermore, in such systems one would expect aphid clones that produce sexual forms early in a year to be at a selective advantage over those that produce them late in a year, which could account for the early production of sexual forms in species such as the lime aphid (Dixon, 1972a). That is, although the three stages in the development of leaves – i.e. growth, maturity and senescence, characteristic of all trees – have a dramatic effect on the reproductive activity of their aphids, it is not the only factor determining the shape of the within-year population profile. A major factor is migration, with isolated mature trees possibly suffering greater losses and showing markedly different population profiles to trees in forests. Moreover, the average abundance of aphids on isolated trees is less than on trees in single-species stands, which supports the theoretical predictions (Dixon & Kindlmann, 1990). Thus, in pondering the reason for the differences in the population profiles and average levels of abundance of aphids on trees, differences in host-plant phenology and isolation appear to be more important than any intrinsic differences in nutritional quality (p. 147).

This tends to indicate that the time scale over which most regulation occurs in deciduous tree-dwelling aphids is considerably less than a year. The density-dependent processes operate mainly on individuals during the course of their development, and the combined effect serves to bring the population density back to an equilibrium trajectory. That is, the density-dependent processes operate continuously rather than between generations. Therefore, analyses of data sets made up of yearly totals of the numbers caught in suction traps will not reveal when aphid abundance is regulated. However, once there is a good understanding of how the numbers of a particular species of aphid on its host plant are regulated, as is the case for the sycamore aphid, then aerial populations can be predicted and the prediction validated by reference to suction-trap catches. The pattern in the yearly catches of the sycamore aphid in the Rothamsted Insect Survey trap at Dundee gives strong support to the contention that the within-year 'seesaw' in abundance is an important feature of the population dynamics of this aphid.

There are studies on the population dynamics of coniferous tree-dwelling aphids. The longest (12 years) is that of Kidd (1990a,b,c) on the large pine aphid *Cinara pinea*. The average numbers of this species recorded each year showed a four- to five-year cycle in abundance. To account for this, Kidd suggested that the natural enemies and/or induced plant defences showed a delayed density-dependent response operating between years. A simulation model developed to explore the effect of these two factors predicted a stable five-year cycle, but only under a very restricted set of conditions. The assumptions regarding the nature and operation of these delayed responses need to be tested. Yamaguchi (1976) gives an account of a 10-year study of the population dynamics of the Todo fir aphid *Cinara todocola*, on young saplings of Todo fir growing in a nursery.

Unfortunately, the interpretation of the results is confounded by changes in the quality of the host plant for this ant-attended aphid, which first increased and then decreased over the period of the observations. However, the rate of population increase in early summer is density dependent. The study by Furuta (1988) on *Cinara tujafilina* over three to five years revealed that, as in the previous species, there is only one peak of abundance each year and that there is a tendency for years with high and low peaks in abundance to alternate. This was attributed to the activity of predators, in particular syrphids. This claim was supported by a predator removal experiment, which resulted in higher peak abundance in two out of three years. Scheurer's (1964, 1971) short-term (three year) but detailed studies on several species of *Cinara* lend support to the idea of a 'seesaw' effect, with high numbers resulting in very few overwintering eggs, and vice versa. A major factor in this appears to be migration, which is more marked in years when aphids are abundant in spring. The consequence of this, as there is predominantly only one peak in abundance each year, is that these species tend to alternate in abundance from year to year. Similarly, a seven-year study of *Cinara pectinatae* revealed that the number of eggs laid in autumn on silver fir *Abies alba* is inversely related to peak number in summer (Maquelin, 1974). Because of its economic importance, the green spruce aphid *Elatobium abietinum* has been extensively studied, but only intensively over relatively short periods at any one site. These studies have revealed direct density-dependence acting on recruitment and migration, and cold weather in winter the major disturbing factor (Hussey, 1952; Day & Crute, 1990). In summary, coniferous and deciduous tree-dwelling aphids show similar population dynamics. In particular, there is some evidence for the 'seesaw' effect in coniferous tree-dwelling aphids. There is a need for the notion that the delayed effects of natural enemies and/or plant defences operate over periods of a year to be tested experimentally.

There is no doubt that the natural enemies of aphids can reduce their rate of increase, occasionally dramatically, however, as indicated in Chapter 6 there is no evidence that natural enemies regulate sycamore aphid abundance. Quite the reverse, it is the abundance of the aphid that determines the abundance of the natural enemies.

For good pragmatic reasons, there are few long-term studies on insects. Such studies need to be encouraged as they provide the reality against which to test theoretical predictions. Population studies at different sites within the range of an aphid should also be encouraged, as there are indications that the within-season dynamics of the sycamore aphid differ in the north and south of the UK (Dixon *et al.*, 1993). In the case of tree-dwelling aphids, population regulation appears to be mainly by means of direct and delayed density-dependent factors operating within a year, with weather and natural enemies perturbing the system. Differences in within-year population profiles appear to be more a consequence of phenology and spatial isolation than nutritional quality of the host plants. Hopefully, this goes some way to account for population cycles regarded as important for the general understanding of population ecology (Turchin, 2001).

In summary, the very big differences that the sycamore and Turkey oak aphids show in within- and between-year dynamics can be accounted for in terms of the intraspecific processes highlighted in Chapter 6. The Turkey oak aphid has essentially only one peak in abundance each year whereas the sycamore aphid has two. This is mainly due to the extent to which the two aphids decline in abundance in summer, which may reflect the abundance and degree of isolation of the two host trees. The cyclical pattern observed in the between-year dynamics in the Turkey oak aphid is a consequence of the inverse relationship between the size of the spring peak and the population rate of increase in autumn, which in the absence of 'noise' in the system results in a more cyclical pattern, with lower amplitude and a weak 'seesaw' effect. The stability in between-year abundance in the sycamore aphid is attributed to the two within-year peaks in abundance being more similar in size and that a change in the size of one peak is compensated for by a similar and opposite change in the other. The cyclical pattern in the yearly suction trap catches of sycamore aphids at Dundee is a consequence of this trap mainly only catching the aphids that migrate in autumn. That is, it mainly only samples one of the peaks in abundance of this aphid each year – the autumn peak.

A simple model that incorporates the effect of cumulative density on population rate of increase and a within-season varying carrying capacity predicts the observed trends in abundance. Thus, in general, their dynamics appear to be explicable in terms of two parameters. The literature also supports the contention that intraspecific competition and the 'seesaw' effect are common features of tree-dwelling aphid biology. A few studies indicate that natural enemies also have a role but this needs to be critically assessed.

8

Risky dispersal

Organisms allocate their resources either to growth or dispersal. It is, therefore, of interest to understand what determines the relative benefits of the two options. The specific case to be addressed in this chapter is dispersal in tree-dwelling aphids. It is an important component of the reproduction of each clone: the evolutionary individual.

For a long time it was speculated (Taylor, 1974a), and now appears that dispersal in aphids is risky, with fewer than 1 in 100 locating a host (Ward *et al.*, 1998). The incidence of dispersal is highly variable, with local populations of some species of tree-dwelling aphid declining by up to 63% per week in late summer mainly as a result of mass emigration, whereas others have low dispersal rates. Host specialists living on long-lived hosts such as trees should stay put because density-dependent effects are rarely likely to reduce replacement to fewer than 1, and only 1 in 100, or less, will survive if they disperse. Thus it is surprising that many species appear to show a marked tendency to emigrate.

Emigrants from one local population could be immigrants to another local population. Therefore in attempting to define the place a species occupies it is important to consider the dispersal and/or migration requirements of the organism in question (Andrewartha & Birch, 1984; Huffaker *et al.*, 1999). If a population is 'a group of individuals of the same species that live together in one area of sufficient size to permit normal dispersal and/or migration behaviour and in which population changes are largely determined by birth and death processes' then 'the emigration and immigration rates become negligible or, at least, roughly balanced' (Berryman, 2002). That is, resolution of the problem may partly depend on how one conceives nature (Andrewartha & Birch, 1984).

THE PROBLEM

Local populations of the sycamore aphid, on mature trees, decline
sharply during summer. The numbers of this aphid in the lower
canopies of eight trees in Glasgow, during the period 1966–74, declined
each summer by between 77% and 93%. These losses are predominantly
due to the disappearance of second generation adults and is termed
'risky dispersal'. It is not the result of their re-colonizing the upper
canopy (p. 29) nor 'natural' mortality through senescence, and inver-
tebrate predators are extremely scarce on sycamore in late summer.
Also, the disappearance rate was unaffected by caging branches with
coarse netting, so predation by birds is unlikely to have been involved
(Dixon, 1969). The remaining explanation, emigration, is supported by
the fact that a suction-trap caught high numbers of migrating aphids
and uninfested saplings were colonized by large numbers of migrants
in those weeks in which the numbers on the trees declined (Dixon,
1969).

How is such behaviour maintained? There are several explanations that might account for the phenomenon of 'risky dispersal'.

1.   *Random extinction.* The death of a tree results in the extinction of
      a local population. If a local population has some probability of
      extinction, e, then the evolutionarily stable dispersal (ESD) rate
      is e, that is, the fraction of each population that disperses each
      generation should equal the fraction of the local population that
      goes extinct.
2.   *Kin competition.* If members of a regulated local population are
      more closely related to each other than to members of other
      populations, selection for the avoidance of kin competition may
      result in dispersal between populations (Hamilton & May, 1977;
      Gandon & Michalakis, 1999; Plantegenest & Kindlmann, 1999).
3.   *Asynchrony.* If conditions in different local populations fluctu-
      ate asynchronously then it will often be beneficial to emigrate
      even from currently low-density populations. Even if density-
      dependent regulation occurs within each local population that
      has asynchronous dynamics, then the dispersal rate should evolve
      towards a rate high enough to synchronize them at which point
      dispersal is selectively neutral (McPeek & Holt, 1992; Holt &
      McPeek, 1996; Doebeli & Ruxton, 1997). Clearly the ESD will be
      lower if there is a cost to dispersal. Dispersal may also be favoured
      if different local populations are subject to negatively correlated

or uncorrelated environmental noise. Fluctuations in fitness in highly suitable habitats through, for example, chaotic dynamics, may even result in selection for dispersal to population sinks, which are maintained by immigration (Holt, 1997).

Despite the wide-ranging and often elegant theoretical developments in this field, few empirical studies have set out to distinguish between the alternative explanations for dispersal. Here an attempt is made to determine the factors maintaining the high incidence of migration in the sycamore aphid. First the migration is described in the context of the aphid's life cycle and its host's phenology, and then the processes and the parameter values expected on theoretical grounds to determine the evolution of dispersal are examined. These theoretical predictions are then related back to the phenomenon of dispersal in the sycamore aphid.

THE SYSTEM

There are three peaks in flight activity during the course of a year (Figure 8.1). Only the summer migration of second-generation adults is associated with a decline in population abundance on the trees. As stated above this decline is not a result of the aphids re-colonizing the upper canopy of the trees, natural mortality or natural enemies (Dixon, 1969). A full explanation should not only shows that migration is adaptive but accounts for at least three of its features: timing, magnitude and dependence on density.

Timing

The first peak in flight activity occurs when the leaves of the trees are still growing and provide a high quality food source, and the aphids have a high reproductive rate (Figure 8.1). At this time there are very big differences in the occupancy of trees by aphids, with the late-breaking trees less heavily infested than the early-breaking trees. As there are clear advantages in terms of the rate and duration of reproduction in being on late-breaking trees this might account for the timing of the first peak in flight activity (cf. Furuta, 1986, 1990a; p. 149). The third peak in flight activity occurs at a time when the leaves of many trees are beginning to senesce. If an adult aphid remains on a tree until leaf fall,

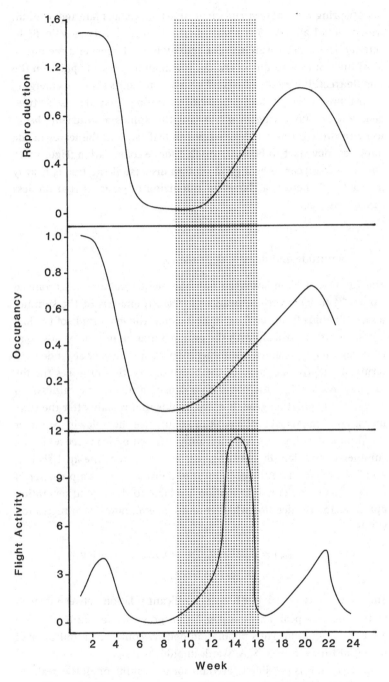

Figure 8.1. The average reproductive rate, occupancy of the upper canopy and flight activity of aphids during the course of a year. The stippled area corresponds to the period when sycamore aphid populations decline in abundance in summer.

her offspring will not have sufficient time to develop, mate and oviposit, which would appear to account for the timing of this peak in flight activity. The second peak in flight activity is the largest and the one of most interest as it results in the decline in abundance of aphids on the sample trees. It consists of second generation adults that are emerging from reproductive diapause. In terms of timing it occurs just as trees generally are improving in quality for the aphids as nutrients begin to be recovered from the leaves prior to leaf fall and the leaves in the upper canopy start to become suitable for re-colonization (Figure 8.1). This would appear to be a good time to disperse if the tree is heavily infested and there is a chance of colonizing other trees that are less heavily infested.

Magnitude and density dependence

The summer decline, averaged over the eight sycamore trees, ranged from 77% to 93%, with a mean of 87.5%. An analysis of the summer losses of aphids from the trees indicates that the magnitude of the loss is consistently associated with the peak population in June or July. That is, as reported in Chapter 6, summer migration is density dependent. In terms of adaptive significance the above observation provokes the following question. Is aphid population density a predictor of fitness? An analysis of population growth of the aphids that remain after the summer migration $[N_{t\,min}]$ to the peak in abundance the following summer $[N_{t+1}]$ indicates that this is significantly affected by year, tree and peak summer population. By omitting year and tree from the analysis it is possible to determine the usefulness of density alone as a predictor of fitness: this is the only information available to the second-generation aphids facing the decision whether or not to emigrate. Linear regression yields:

$$\log [N_{t+1}/N_{t\,min}] = 2.081 \pm 0.375 - (0.523 \pm 0.155) \log [N_t],$$

$$F_{1,62} = 11.38, P = 0.001. \tag{8.1}$$

That is, there is a weak but highly significant relation between density at the summer peak and the population growth rate $(\log [N_{t+1}/N_{t\,min}])$ after the migration. Thus population density could be an indicator of potential future fitness (S. A. Ward, unpublished data).

That is, it is possible to account for the timing of all the peaks in migration, and in particular in the timing of the summer migration, its magnitude and dependence on density.

EXTINCTION AND KIN RECOGNITION

The only event likely to cause the extinction of an established sycamore aphid population is the death of the host. It is important here to confine the estimate to mortality of *mature* trees, rather than saplings, whose bark is too smooth to provide oviposition sites, and which thus cannot give rise to significant aphid populations in spring.

The available data suggest that once a sycamore has reached 5–10 years it is likely to survive another 50–100. Aphid populations were monitored on eight mature trees in Glasgow for eight years, and by Simon Leather on 50 in Silwood Park, Berkshire, for 10 years. In neither study did any hosts die – there was no mortality in 564 tree-years. Thus a very conservative estimate of the life span of sycamore is 10–100 years, giving an extinction rate (e) of between 0.1 and 0.01. The estimates are 'conservative' because they are likely to yield predictions of a higher migration rate than if a longer life span is assumed. A second reason is that it is assumed that e does not vary between years. If it does, the benefits of migration are reduced.

Wynne *et al.* (1994) studied genetic variation in second-generation sycamore aphid adults at four spatial scales. They found that genotype frequencies were homogeneous at the level of leaves, branches, trees and regions in the south of the UK. They also found no significant deviation from Hardy–Weinberg equilibrium. This implies that inbreeding is negligible and the autumnal populations on individual trees are made up of many clones. Within-tree relatedness may increase slightly as a result of variation in the fitnesses of the sexual females and first generation individuals, and survival of second-generation nymphs, but will still remain low.

An important related point is that Wynne *et al.*'s data provide no evidence for any associations among clone-mates: rare genotypes were rare everywhere – they were never found clumped on a single leaf or branch. The above is not unexpected bearing in mind the high incidence of migration and movement between leaves.

Synthesis

Gandon & Michalakis (1999) have synthesized the work on these two factors, providing a single equation for the ESD as a function of the cost of dispersal, c, the extinction rate, e, and the within-population relatedness, R:

$$d^* = \frac{A - \sqrt{A^2 - 4eB}}{2B} \tag{8.2}$$

where

$$A = c + (1 - c)e^2 + e - R(1 - e)$$
$$B = [c + e(1 - c)]^2 - R(1 - e)$$

Table 8.1. *Evolutionarily stable dispersal rate, predicted using equation 8.2*

Relatedness, R	Cost of dispersal, c	Extinction, e	ESD, d*
0	0.99	0.1	0.101
		0.01	0.010
	0.9	0.1	0.110
		0.01	0.011
0.1	0.99	0.1	0.111
		0.01	0.011
	0.9	0.1	0.122
		0.01	0.012

(After S. A. Ward, unpublished data.)
ESD, Evolutionarily stable dispersal.

Table 8.1 shows the ESD rate, d^*, calculated using equation 8.2 for a realistic range of parameters. Over these ranges, R and c have trivial effects; what determines d^* is the extinction rate. Now, even with the most conservative assumptions – that an established sycamore lives for only 10 years, that 10% of dispersers find hosts and that sycamore aphids normally live surrounded by their full sibs – the combination of kin competition and host mortality cannot account for the mass emigration observed in summer.

A second reason why the Hamilton–May explanation must be discarded concerns the spatial scale of the mechanism of competition. Since all the evidence indicates that competition occurs only between adjacent aphids on a leaf, a sycamore aphid in a crowd of clone-mates could avoid kin competition simply by moving to another leaf, away from its kin, rather than embarking on a hazardous migration to another host.

ASYNCHRONY

The degree of asynchrony of populations' dynamics clearly depends on the spatial scale of the analysis. Two temporal forms of dispersal are distinguished, the evolution of which will depend on details of the asynchrony: redistribution and alternation.

Between regions

If populations in different regions fluctuate asynchronously, because of, for example, localized environmental fluctuations, then it may be beneficial to a lineage to redistribute itself among regions. If the asynchrony is seasonal, for example, conditions are lush in spring in one area and in autumn in another, then selection for 'region alternation' might lead to the evolution of dispersal.

Between mature hosts

Similar possibilities arise for differences among mature hosts. Depending on their phenology and situation, trees may not all be nutritionally rich at the same time of year, or in the same years.

Between host age-classes

Sycamore populations are not composed of a single age-class. Seedlings and saplings will generally be less well lit than mature trees, so their phenology – timing of bud-burst, leaf expansion, senescence and leaf-fall – may differ. There may thus be seasonal asynchronies in the nutritional qualities of hosts of different ages, so clones might benefit from 'age class alternation'. Apart from their nutritional status, the *abundance* of seedlings and young saplings may fluctuate more than that of more advanced saplings and mature trees. The 'mast year' phenomenon documented for many forest trees may influence the optimal dispersal rate of the aphids that feed on them, since it means that the benefits of dispersing in search of seedlings/saplings are unpredictable.

Regional redistribution

Most aphid migration is thought to cover a range of 20–50 km (Loxdale et al., 1993), so migrants are very unlikely to move to a region where conditions differ markedly from those from which they have dispersed. Mercer (1979) examined the annual catch of sycamore aphids in 12 of the 12.2-m suction traps run by the Rothamsted Insect Survey of the UK. He found strong positive correlations between traps as far apart as 690 km and the correlation between the catches from Rothamsted and Newcastle, 370 km to the north, is highly significant, 0.67 (Figure 8.2).

All the traps, other than that at Reading, are more than 50 km from Rothamsted in Hertfordshire. Although the distances between Rothamsted and the other trap sites are mostly in excess of 50 km,

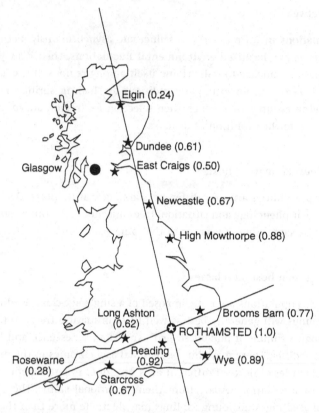

Figure 8.2. The location of some of the Rothamsted Insect Survey suction traps in the UK and the correlation coefficients between the number of sycamore aphids caught by these traps and the Rothamsted trap.

some considerably so, nevertheless there are good positive correlations between the catches at these sites. That is, there are years when the catches are generally good or bad throughout the UK. Similarly, migration maps (Taylor, 1974b) for the sycamore aphid, plotted at four-weekly intervals for the period 1969–73 also indicate that the sycamore aphid tends to be either abundant or uncommon generally throughout the UK in any one year. In addition, the aphid is caught first in the south in most years, which is associated with bud burst occurring earlier in the south than in the north. That is, within the supposed migratory ambit (50 km) of the sycamore aphid, the temporal changes in the aerial population are similar. Thus there would appear to be no advantage in the aphid redistributing itself temporally or spatially between regions.

Redistribution between established hosts

The annual increase in abundance, from immediately post-migration to the following summer's peak, for each of the eight trees in each of the eight years is discussed on page 126. Assuming that these eight trees are representative of the population as a whole it is possible to predict the growth of an hypothetical clone with a set dispersal rate, d, assuming that dispersers are spread evenly among trees. 'Fitness' is estimated as the geometric mean annual increase. If all dispersers survive, the optimal behaviour is for the whole clone to migrate each year. For lower survivorship, fitness initially declines with d, but then rises again, so that if more than about 70% survive, the optimum is still complete dispersal. For survivorship lower than 65%, the fitness curve never rises above the level for a completely sedentary population. That is, although there are big differences in the fitness of the aphid on the eight trees the apparent cost of dispersal would appear to make dispersal disadvantageous (S. A. Ward, unpublished data). This is discussed in more detail below, in particular, whether the eight trees are representative of the population as a whole (p. 133).

Within- and between-year fitness

Differences between trees in the time they burst their buds and shed their leaves are important in determining the survival of resident sycamore aphids from autumn to the following spring (pp. 67–8). Trees that break their buds early consistently do so from year to year and vice versa. Similarly, those that shed their leaves early or late do so consistently from year to year. For the eight sycamore trees studied in detail there was no correlation between the average date of bud burst and that of leaf fall over a period of eight years ($r = 0.16$, $n = 8$, NS). Although eight trees is a very small sample, less detailed observations tend to indicate that the lack of a correlation between bud burst and leaf fall is a general phenomenon. Within the sample of eight trees there is one tree that both breaks its buds early and sheds its leaves late and another on the same site that does the reverse: breaks its buds late and sheds its leaves early. A very high percentage of the aphids that hatch in spring survive to colonize the leaves on the tree that breaks its buds early. In addition, the egg laying females are able to build up in numbers and complete egg laying before the leaves are shed in the autumn. The reverse is true of the tree that breaks its buds late and sheds its leaves early (Figure 6.5).

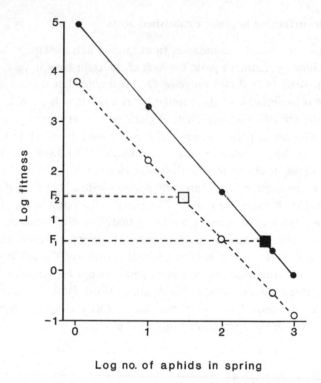

Figure 8.3. Relationships between logarithm of the within-year fitness and logarithm of spring peak numbers for two trees on one site. (F1 and F2 are the average within-year fitness of the aphids on the two trees.)

On all eight trees there is an inverse relationship between peak numbers of aphids in spring and those present in autumn (pp. 70, 84–6). Knowing the average peak number of aphids present in spring on each of the trees it is possible using these relationships to calculate the average autumn peak numbers on these trees (Figure 8.3). These two values for each tree can be used to calculate the numbers present in autumn relative to those present in spring. This measure of fitness on the eight trees varies from 2.5 to 30.4. One can make the same calculations for between-years (autumn to spring) for which the fitness varies from 0.54 to 3.2. Thus, in both autumn and spring it is more advantageous for aphids to be on certain trees. For aphids on the tree that burst its buds early and shed its leaves late the within-year fitness was 3.5 and the between-year fitness 2.4, whereas for those on the tree that burst its buds late and shed its leaves early it was 30.4 and 0.6, respectively. That is, depending on the costs of moving between these two trees, it would on average be more (x9) advantageous to be on one tree in spring and the other tree (x4) in autumn.

There is a strong inverse relationship between the average peak number on each of the trees in spring and the average fitness of the aphids on those trees. Why should this be? The small data set of eight trees, on three sites that experienced different conditions, especially temperature, makes further comparison difficult. However, the two trees with the most strikingly different phenologies were on the same site, so are likely to have experienced similar temperatures. The abundance of aphids on the tree that broke its buds late and shed its leaves early was on average half that on the tree that broke its buds early and shed its leave late (Figure 6.3). Associated with this was a difference in the average fecundity of the aphids. Those on the tree with the lowest aphid abundance were 1.4 times more fecund than those on the tree where they were most abundant (Figure 6.4), although the average weights of the aphids on the two trees did not differ significantly. The aphids on the tree with the lowest average aphid abundance had a higher fecundity in terms of nymphs per adult per day than those on the tree with the highest average aphid abundance, except for one year (1973) when the aphids were equally abundant on both trees (Figure 6.3). That is, the high performance/fitness within years of aphids on one of the two trees is associated with high fecundity and low population density. Clearly, other factors like rate of emigration and immigration also affect the means fitness used here. However, in terms of fecundity there is an advantage in aphids being on trees with the lowest numbers of aphids.

The differences in the fitness of the aphids on the eight trees between years appears to be mainly associated with tree phenology and is unlikely to have been a consequence of movement between trees. Again it would appear to be more advantageous for the aphid to overwinter on certain trees, in particular those that break their buds early and shed their leaves late. However, because such trees are likely to be heavily infested there are advantages in being on other trees for the rest of the year.

Leather's data set

Analysis of the relationship between autumn peak numbers and the numbers present on the leaves the following spring for a larger number (50) of sycamore trees sampled by Simon Leather in Silwood Park, Berkshire, indicates that the aphids on large trees are generally fitter (x10) than those on small trees. The trees varied in size from 1.8 to 20 m. Another feature this analysis revealed was that the occupancy of the small trees by aphids is less and varied more than that of the large trees, with the aphids often becoming extinct on the small trees. That is, the age of a tree could be an important factor determining its quality as an overwintering refuge for aphids. A factor that could be important in this is the number of egg laying sites; the bark of large trees has more crevices, which is the preferred oviposition site of sycamore aphids (p. 21). In addition, small trees are likely to be more shallowly rooted and in dry summers subject to water stress, which leads to them shedding their leaves early. Whatever the reason(s) for the greater fitness of the aphid on the larger trees and the greater variability in the occupancy of small trees the net effect is that the degree of heterogeneity between trees in terms of their quality for aphids both between and within years is much greater than that indicated by the study of the performance of the aphid on the eight trees at Glasgow. It is also likely that the only indication an aphid has of its potential fitness on a tree is the abundance of aphids on that tree.

Asynchrony as an explanation of dispersal

Providing the advantages outweigh the costs of dispersal between trees then if on a tree where aphids are abundant they should disperse and colonize trees where aphids are uncommon or absent. This combined with the tendency for the aphid to fly higher in autumn than in spring (p. 111) increases the probability of the aphids colonizing large trees in autumn, and their tendency in autumn to prefer mature to senescent leaves (Dixon & Mercer, 1983) in them accumulating on those tall trees that senesce last. However, if the cost of dispersal is as high as indicated above then as the first general analysis revealed the difference between the best and worst trees, in terms of aphid fitness within and between years, appears to be insufficient to offset the cost of dispersal. The degree of asynchrony in phenology, abundance and age distribution of the trees will all affect the outcome. However, it is difficult to visualize a situation in this system in which the benefits of dispersal outweigh the apparent costs.

All three explanations are based on advantages that do not compensate for only 1 in 100 aphids surviving dispersal. It is possible that all three explanations apply to some degree and it is the combined effect that is important. However, this seems unlikely. Therefore, is there a need for a paradigm shift in our approach to the problem of risky dispersal? Is it not possible that the success in moving between different aged/sized adjacent trees is considerably better than 1 in 100, especially when the host plant is abundant?

Of the three, the asynchrony explanation has at least some support. It is also an attractive explanation for an entirely different reason. A major problem in aphid biology is accounting for the evolution of host alternation. Approximately 10% of aphids move seasonally between different species of plants, with most spending autumn, winter and spring on a woody plant and summer on an herbaceous plant. Extensive seasonal movement between plants of the same species, as implied by the asynchrony hypothesis and supported by some field evidence, may possibly have facilitated the switch to colonizing a species of herbaceous plant in summer. In addition, the study of Furuta (1986, 1990a; p. 149) lends strong support to the idea that other deciduous tree-dwelling aphids also move seasonally between trees exhibiting different phenologies. However, for this strategy to be successful the host plant has to be abundant and phenologically diverse. In the case of *Kaltenbachiella japonica*, which produces reluctant migrants (Komatsu & Akimoto, 1995), the prediction is that its host tree is rare and/or phenologically uniform.

In summary, there are three explanations that may account for the marked summer exodus of sycamore aphids from mature sycamore trees: random extinction of the host, kin competition between aphids and asynchrony in the quality of the host for the aphid. Random extinction and kin competition can be ruled out. Although it is currently not possible to prove formally that the benefits of moving between trees of markedly different phenologies and ages outweighs the costs, there is some evidence that tends to indicate that asynchrony between host trees could account for the phenomenon of 'risky dispersal' in the sycamore aphid, especially as the host plant is abundant.

9

Seasonal sex allocation

Most tree-dwelling aphids have an holocyclic life cycle, in which there is a series of parthenogenetic generations culminating in the production of males and sexual females, which lay the eggs. In temperate regions egg-laying coincides with the onset of winter, but in other parts of the world it coincides with the onset of a hot/dry season. The switch to sex should be postponed as long as possible as it ends the phase of rapid parthenogenetic reproduction. However, mating and oviposition must occur before the end of the favourable season, which for aphids living on deciduous trees in temperate regions is leaf fall. In the sycamore aphid the production of sexual females is cued mainly by short day-length and low temperature (Dixon, 1971d). That is, in common with other aphids, it shows environmental sex determination. In terms of physiology it is likely that environmental cues trigger neuroendocrine changes in the aphid, which control the sex of its offspring at the time of the maturation division of the ova. The result is that the males and females are present for a relatively short period in autumn at a time when population abundance is changing very rapidly (Figure 9.1). That is, both sexes make only a brief appearance and neither achieve an equilibrium density. In addition, the fitness of a male depends on the *number* of females it fertilizes, while that of a female depends on her *probability* of mating.

The fact that aphid populations are made up of clones increases the possibility of females mating with their male relatives. If local mate competition (LMC) occurs then there will be selection for a female-biased sex allocation (Hamilton, 1967). This is particularly so for species, like the sycamore aphid, which live on the same species of host plant throughout a year (Dixon, 1998). The chance of selfing in these aphids depends on how many clones are present on an individual host at the time of mating. In the extreme case of one clone resulting in sib mating,

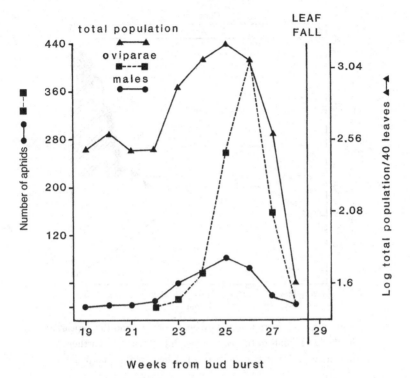

Figure 9.1. Changes in the abundance of aphids that occur on the leaves of sycamore in autumn prior to leaf fall. The trends are for the logarithm of the total number of aphids, and the actual numbers of adult males and sexual females (oviparae).

a parent can maximize her fitness by producing many daughters and just enough sons to mate with them. In addition, the fitness of males and females may depend in different ways on environmental variables. In those cases where there is environmental sex determination, as in aphids, the skew in the primary sex ratio depends on the form of the relationship between fitness and environment (Charnov & Bull, 1989a). Existing models assume that the sex ratio is not only at evolutionary but also demographic equilibrium. In organisms with brief mating seasons like tree-dwelling aphids, however, this is not true (Figure 9.1).

THE SYSTEM

Deciduous tree-dwelling aphids live in a seasonal environment, which is only suitable for reproduction up to the time of leaf fall. In those species

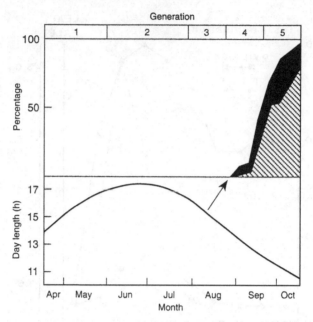

Figure 9.2. The relationship between day-length and the appearance in
the field of sexuals of the sycamore aphid. Open area, parthenogenetic
individuals; solid area, males; cross-hatched area, sexual females.

of aphids that live on evergreen trees, low temperature is likely to be
the factor that brings the favourable season to an end. That is, there
is some time by which aphids should complete the laying of eggs. This
presents aphid clones with an optimization problem: when and how to
switch to sexual reproduction in order to maximize both the production of viviparous parthenogentic and sexual individuals. They appear
to have resolved this problem in part at least because there is a high
degree of synchronization between the appearance of sexual forms and
the onset of senescence of the leaves of sycamore. This poses the following questions: (a) What mechanism(s) do aphids use to synchronize
their phenology with that of their host?; and (b) What is the adaptive
significance of the patterns in time in the allocation of resources to
male and female production?

Mechanism

Although many aphids hatch from eggs early in the year and are present
in spring when days are as short as they are in autumn, the sexuals
rarely appear in the field before autumn (Figure 9.2). This is because

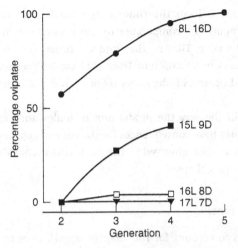

Figure 9.3. Progressive changes in the percentage of sexual females (oviparae) produced by successive generations of the sycamore aphid reared from egg hatch in one of a range of day-lengths. Generation 1 is the fundatrix; L, hours of light; D, hours of darkness.

of the operation of an intrinsic timing mechanism referred to as an 'interval timer' (Lees, 1960, 1966).

In the sycamore aphid the interval timer mechanism control-ling sexual female production does not function independently of the environment while it is 'running out'. With the passage of time the ratio of the number of sexual to parthenogenetic females produced in successive generations, reared under constant short-day conditions, gradually increases and does so more rapidly in the shorter photoperi-ods (Figure 9.3). That is, aphids present in spring can produce sexual females but only under day-lengths shorter than those prevailing in spring. However, the latter generations are sensitive to the short day-lengths similar to those prevailing in spring and autumn. In this way the production of sexual females is synchronized with the onset of autumn. In addition, the incomplete inhibition of the interval timer in autumn enables clones to continue to produce some parthenogenetic individuals in autumn when the senescent foliage of sycamore pro-vides a rich source of food. This is advantageous when aphids are on those trees that in any one year shed their leaves late and/or present in autumns when leaf fall is late (Dixon, 1971d).

In contrast males do not appear until the fourth generation, and their appearance does not seem to be related to day-length (Dixon, 1971d). Sycamore aphids reared continuously under long-day conditions (17 h) at 15 °C from the time of egg hatch produce, in autumn, males in proportions comparable to those observed in the field, but no sexual females. That is, the 'interval timer' controlling male production appears to be different from that controlling sexual female production and operates independently of both day-length and temperature.

The difference in the way the production of males and sexual females is controlled may have consequences for the time of appearance of the sexes and the sex ratios observed in the field. This is considered in more detail later in this chapter.

Adaptive significance

A full explanation has to account for the adaptive significance of two features in particular: (1) the female-biased sex ratio; and (2) the tendency for males to appear before the sexual females.

The sex ratio varies between years and trees, with the average for the eight sycamore trees ranging from 1.4 to 3 times more sexual females than males observed. The trees on any one site with the highest female bias were those that supported the highest population densities of aphids. However, the variation from year to year on any one tree does not appear to be associated with the variation in population density from year to year. That is, the consistent feature is that the sex ratio is biased in favour of females. However, this bias is stronger on some trees and in some years. The birch aphid *Euceraphis betulae*, has a similar female-biased sex ratio to the sycamore aphid: two to three times more sexual females than males (Daag, 2002). In other species the bias in favour of females can be greater, for example, in the alder aphid *Pterocallis alni* (De Geer) it is 1:5 (Gange, 1985), in the lime aphid 1:7, and in the Turkey oak aphid 1:13. There is also some indication that the lower the female bias in sex ratio the more time the males spend mate guarding (Dixon, 1998; Daag, 2002).

Males started to mature before the sexual females (protandry) in 41 out of the 63 tree years (Figure 9.1). That is, protandry was observed on all trees and in all years. The degree of protandry, however, appears to be associated with the abundance of aphids in autumn. When aphids are abundant some males are recorded as much as four weeks before the first sexual female and when aphids are uncommon females may be recorded before the first males (see Figure 9.6). That is, the degree of protandry appears to be associated with how well the aphids do in autumn.

Theory

Sex-allocation theory concerns the evolution of patterns of investment in reproduction through male and female pathways (Charnov, 1982). Fisher's (1930) theory predicts that natural selection should drive populations to an equilibrium at which half of the parental resources are allocated to sons and half to daughters. Three processes have since been suggested to lead to biased sex-allocation ratios. First, if females mate with relatives, then LMC results in selection for female-biased sex allocation (Hamilton, 1967); in the extreme case – sib mating – a parent can maximize the number of its grandchildren by producing many daughters and just enough sons to mate with them. Second, in social insects, only the females contribute to nest construction and the rearing of brood: this sex-dependent 'local resource enhancement' means that sex allocation should be female-biased. Third, the fitness of males and females may depend in different ways on age or environmental variables. If females benefit more than males from being large, then the evolutionarily stable strategy (ESS) is for individuals to be male if they are small and female if they are large. Under sex change and environmental sex determination the skew in the primary sex ratio depends on the forms of the relations between fitness and age (or environment) (Charnov & Bull, 1989a,b).

Most models of sex allocation assume not only evolutionary but demographic equilibrium. This need not be true, however, if the mating season is brief, or if the age-dependent survival and mating rates differ between the sexes. Werren & Charnov (1978) and Seger (1983) examined sex allocation under partial bivoltinism. If some males from the first generation survive to mate with females from the second, then the ESS is a male-biased allocation in the first generation and a female bias in the second. This is similar to the system prevailing in the sycamore aphid in autumn. Reproduction is continuous and the generations overlap. Seamus Ward (unpublished data) applied Werren & Charnov and Seger's concept to a large number of time intervals in which sex allocation can be optimized. He also incorporated asymmetries that result in the two sexes having different time-dependent fitnesses: a female may be able to fertilize her full complement of eggs using sperm from a single mating, whereas a male's fitness is proportional to the number

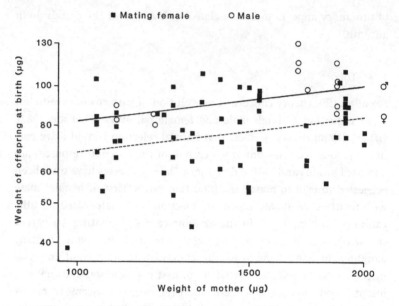

Figure 9.4. Relationships for the sycamore aphid between weight of sexual female (oviparae) and males at birth and the weight of their mothers.

of females he mates. In addition, the fact that mothers invest more, in terms of mass, in male (x1.24) than female offspring at birth (Figure 9.4) was taken into consideration.

However, before considering the predictions of Ward's ESS model it is necessary to consider how important LMC is likely to be in the sycamore aphid. In most species of tree-dwelling aphids, including the sycamore aphid, the sexual females are wingless. The chance of inbreeding or, more precisely, clonal self-fertilization depends on how many clones are present on a plant at the time of mating. On herbaceous plants this may be small, so that selfing is common; on trees the number of clones is likely to be large, and selfing correspondingly rare. This is particularly true of the sycamore aphid, which compared to other tree-dwelling aphids shows very high levels of dispersal (Chapter 8), and consequent mixing of clones. A genetic analysis of the sycamore aphid's population structure confirms that the populations on individual trees are indeed made up of many clones (Wynne et al., 1994). That is, LMC is unlikely to be an important factor determining the female-biased sex ratios observed in the sycamore aphid.

Ward tested the predictions of his ESS model by comparing them with what was observed in the field. The precise predictions depend on

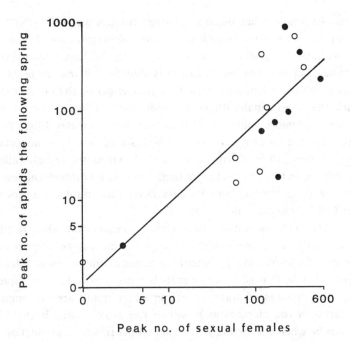

Figure 9.5. Relationship between peak number of aphids colonizing leaves in spring and the peak number of sexual females the previous year for two trees on the same site over a period of eight years.

whether females mate repeatedly and fertilize each ovum with sperm from the most recent mating or mate only once, and on the rates of search, and of male and female mortality. As the slope of the relationship between the logarithm of the peak numbers in autumn and that of the following spring is not significantly different from 1 (Figure 9.5) there is no evidence that sexual females are less successful at attracting mates when population density is low. For the male there is some evidence that competition for mates increases as population density increases. Over a two-year period, the ratio of males to virginoparae caught in the suction trap relative to those on the leaves of two adjacent trees was significantly six times higher in the year when population density was highest ($\chi^2 = 128$, $P < 0.001$). That is, the tendency of males to disperse is greatest when population density is high.

That the fitness of a male depends on the number of females it fertilizes, while that of a female depends on her probability of mating, has two important consequences. First, if the males are long lived, then there are advantages in them maturing early – protandry. The length of life of unmated males of the sycamore aphid is shorter than that of unmated females when caged on saplings in a greenhouse (Wade, 1999). This contrasts with the observation on the willow carrot aphid *Cavariella aegopodii* (Scopoli), which indicates that males live longer than sexual females, but that the length of life is shortened if they mate. Males caged with many females survive for a shorter period than those caged with one or no females (Dixon & Kundu, 1997). Whether the survival of male sycamore aphids relative to that of the females differs from that recorded for *Cavariella* needs to be resolved as it is an important issue. Second, there should be a seasonal bias in the sex allocation, depending on how the total reproductive rate varies through the season and mainly on the longevity of males. Overall this results in a variable but female-biased sex allocation.

The field observations show clear protandry only when population density in autumn is high (Figure 9.6). This may be adaptive and associated with the low probability of males mating when population density is low. That is, finding a mate is more important than improving one's chances of multiple mating when aphids are uncommon. In terms of the mechanism by which this is controlled it could be driven by when the second generation adults resume reproduction in autumn. In Glasgow, high populations in spring are usually followed by a long reproductive aestivation of the second generation adults and low autumn populations. That is, there is a delay in reproduction and more importantly in the maturation of the fourth-generation individuals. In the south, where abundance is generally much lower, the delay in coming out of reproductive aestivation may be due to the generally warmer and longer summers in the south than the north. That is, the degree of protandry is associated with when the aphid comes out of aestivation. This is supported by a significant correlation between synchrony in the development of the sexual morphs and the incidence of reproduction in July. When large numbers are born in July, males tend to occur earlier than sexual females (Wellings *et al.*, 1985).

Both in the north and the south of the UK, the sycamore aphid has, overall, a female-biased sex ratio. In the north it varies from year to year and between trees, ranging from 0.8 to 3 times more females than males. Both the female bias and the degree of protandry also appear to be positively associated, significantly so, on some trees. As males are winged and more mobile than the unwinged sexual females and even more flight active than the winged parthenogenetic females, males may move between trees. Thus, the sex ratio on any particular tree could be affected by the movement of males between trees, and

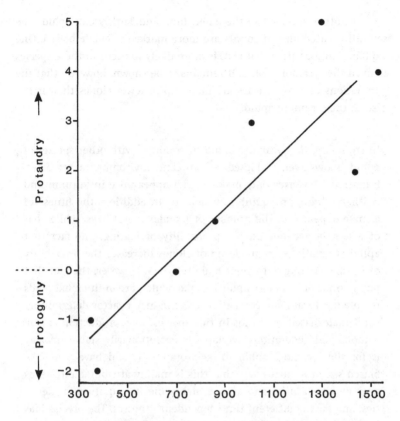

Figure 9.6. The relationship between the time of maturation of the first males relative to that of the first females in weeks and the peak autumnal abundance of aphids on two trees on the same site over a period of eight years.

their tending to accumulate on those trees that shed their leaves last in any one year. In addition, mature sexual females spend some time each day laying eggs on the trunk and main branches of their host trees (Wellings, 1980) and are therefore not included in the sample counts of the aphids on the leaves. That is, sex ratios based on field counts of the aphids on leaves are unlikely to be accurate. However, rearing this aphid on saplings under near natural conditions in the field in autumn (Wade, 1999) and under short-day conditions in the laboratory (Dixon, 1971d) also indicates that the adult sex ratio is female biased. Therefore, although the precise operational sex ratio in the field is uncertain the data indicates it is variable and female biased.

In other species like the alder, lime and Turkey oak aphids the sex ratios based on leaf counts are more markedly female biased. One explanation for this is that LMC is more likely to occur in these species than in the sycamore aphid. It remains to be shown, however, that the populations of these species are made up of fewer clones than is the case in the sycamore aphid.

In summary, the sycamore aphid in common with other species of aphids shows a female-biased sex ratio and environmental sex determination. The sexes only make a brief appearance in autumn and neither achieve an equilibrium density. In addition, the fitness of a male depends on the number of females it fertilizes, while that of a female depends on her probability of mating. The fact that aphid populations are made up of clones increases the probability of females mating with their male relatives. However, the fact that populations of sycamore aphids, in particular, are multiclonal tends to rule out local mate competition as the major factor determining the female-biased sex ratios in this species. As reproduction is continuous and generations overlap the evolutionarily stable strategy is for the sycamore aphid to be protandrous and have a female-biased sex ratio. It is likely that this is mainly attributable to sexes only making a brief appearance, not achieving equilibrium densities, and having different time dependent fitnesses. The precise bias is likely to depend on environmental factors like population density and time of leaf fall. Our current understanding of the ecology of sex in tree-dwelling aphids is likely to be greatly improved by more extensive ecological and detailed genetic and theoretical studies of the problem. However, the realization that the marked seasonality of the habitat occupied by these aphids has played a major role in shaping sex allocation is an important advance. It is now up to others to build upon this work.

10

Aphids and tree fitness

Because they are small, and most do not damage the foliage of trees, aphids are not usually thought of as adversely affecting tree growth and reproduction. However, what they lack in size they make up for in abundance. The 116 000 leaves of a 20 m sycamore tree can be infested with as many as 2.25 million aphids, equivalent in mass to a large rabbit (Dixon, 1998), and the 58 000 leaves of a 12 m lime tree with 1.1 million aphids, equivalent to six great tits or four sparrows.

The studies considered here, involving approximately 150 tree years, revealed marked intraspecific differences between trees in the average level of infestation by aphids. However, in all cases these differences could be attributed mainly to differences in phenology between trees. That is, none of the trees appeared to show any antibiotic resistance to aphids. However, it must be stressed the identification of resistant trees was not a specific objective of this study.

AUTUMN LEAF COLOURS: SIGNAL OR SCREEN

Sycamore is a maple and in common with other species of *Acer* can have brightly coloured leaves in some autumns. This variation in sycamore is attributed in part to aphids, with high levels of infestation in spring resulting in leaves falling green, whereas low levels result in them turning yellow before leaf fall (Dixon, 1971a). Archetti (2000) and Hamilton and Brown (2001), however, suggest that the red and yellow leaf colours displayed in certain parts of the world signal a tree's ability to defend itself against insects. That is, the bright autumn coloration serves to signal defensive commitment against autumn colonizing insects. Better defended trees have brighter autumn leaf coloration. If this is so then at an interspecific level, tree species suffering greater insect attacks should invest more in plant defence and defence signalling. On an intraspecific level, within signalling species, the most defensively committed individuals should produce the most intense displays (Hamilton & Brown, 2001). Hamilton and Brown addressed the interspecific test using the data available in the literature. This revealed that autumn coloration, particularly yellowness, is more intense in tree species that have a high aphid species richness and that the numbers of specialist aphid species correlate most strongly with leaf colour.

Both Archetti (2000) and Hamilton and Brown (2001) propose that the handicap theory (Zahavi, 1975; Grafen, 1990) could help explain how autumn leaf colours act as a reliable indicator of plant defence. It involves a significant cost, which only vigorous and therefore well-defended plants could afford to incur. In the absence of any other explanation for autumn leaf colours this proved to be an attractive and popular idea.

Of the phytophagous taxa that might respond to such signalling, aphids were highlighted by Archetti (2000), and supplied the supportive correlative evidence presented by Hamilton and Brown (2001). As well as ignoring the marked effect of density dependence on aphid fitness (Chapters 6 & 8), there are several other aspects of aphid ecology, however, that are at odds with Hamilton and Brown's theory.

Timing of the signal

Of the two groups of aphids that feed on trees the host-alternating species return to trees in autumn *before the leaves change colour* and depart from the trees in late spring or early summer (Dixon, 1971c). The primary reason for the early colonization is that in most cases their offspring have to complete their development and lay the overwintering eggs *before* leaf fall (Ward *et al.*, 1984; Dixon, 1998). Therefore, they are likely to be strongly selected to respond to cues that indicate time to leaf fall – the ecological deadline in this system. The non-host alternating species, like the sycamore aphid, colonize trees throughout the year, with a high proportion of the colonization by this group occurring in spring and summer (p. 124). Given that most host plants are located well before leaf senescence by both groups of aphids, it is unlikely that the trees would be selected to advertise their defensive capability *after* the key host-finding phase has occurred. It could be argued that even if the advertisement catches only a proportion of the potential population, then it might be worth signalling. However, this raises an important question, why has there not been selection for a better timed signal?

Effect of environmental factors on autumnal coloration

Not only does the autumn foliage of a tree vary in colour from year to year (Dixon, 1971a) and tree to tree, but also from one leaf to another on the same branch (Chang *et al.*, 1989). This greatly complicates the assessment of the defensive ability of a plant. Why should the signal from a plant be so variable if it is primarily for conveying information about the plants defensive capability?

A key element of Hamilton and Brown's (2001) hypothesis is that the aphid response is related not to the colour per se but rather to the relative intensity of colour. Therefore, trees are in competition with each other to produce the brightest colour in order to avoid aphid attack (Archetti, 2000). If environmental factors, which are not directly correlated with the likelihood of aphid colonization, influence the intensity of leaf colour, then clearly this weakens the signalling theory. There is mounting evidence to suggest that leaf coloration is environmentally regulated, with trees growing under harsh conditions being the most colourful. That is, it runs counter to the hypothesis in that trees that experience harsh conditions and likely to be defensively weakened tend to have the most colourful leaves.

The study of Furuta (1986, 1990a) is also revealing in this respect, and Hamilton and Brown (2001) cite his work in support of their hypothesis: autumnal migrants of *Periphyllus californiensis* prefer to colonize *Acer palmatum* with yellow-orange leaves over those with red leaves. In addition, they also cite another observation of Furuta: in spring the fecundity of the offspring of those aphids that colonize the trees with the reddest foliage is less than that of those that colonize the trees with yellow leaves. However, Hamilton and Brown ignore the fact that the trees studied by Furuta were not all equally exposed to the sun. The autumnal colour of those in the shade is yellow-orange and those in the sun red. More importantly for the aphid the trees in the shade shed their leaves later and burst their buds earlier than those in the sun. By colonizing trees with yellow-orange foliage in autumn this aphid is more likely to produce overwintering eggs and complete an extra generation the following spring than those that colonize trees with red foliage. That is, by avoiding trees with red foliage this aphid is likely to be fitter, which is in accordance with Hamilton and Brown's hypothesis. However, what happens in late spring/early summer is equally important. The late-breaking trees, those with red foliage the previous autumn, have growing leaves, which are very attractive to aphids, at the time when the leaves of the early-breaking trees are maturing and the emigrants produced on these trees then colonize the leaves of the late-breaking trees. That is, the trees with red foliage in autumn are likely to be more heavily infested with aphids in late spring/early summer than the trees that had yellow-orange foliage the previous autumn. By dispersing between trees the aphid exploits the spatial heterogeneity in plant quality and colonizes those trees that are the most suitable for reproduction at a particular time (Furuta & Sakamoto, 1984).

Viewed in terms of the whole season, rather than just the autumn, this aphid appears to respond to *cues*, which enable it to occupy at any particular time the most favourable part of what is a very heterogenous/coarse-grained environment. By being yellow-orange rather than red in autumn does not signal a lack of defensive commitment to the aphid, but that the tree's phenology is more closely synchronized with its life-cycle requirements. At other times, another part of the environment is more favourable, and in late spring/early summer it is the trees that have red foliage in autumn that are preferred.

Correlative evidence

It is widely reported that autumn leaf colour of individual tree species tends to be more intense at high latitudes. For instance, recently Hoch *et al.* (2001) demonstrated that anthocyanin accumulation in deciduous species of nine genera is correlated with geographical origin. Species from the relatively mild climates of Europe do not display the 'high anthocyanin production' characteristic of those from the continental climate of northern USA and Canada. As such comparisons involve many congeneric trees a phylogenetic explanation appears unlikely. In addition plant species diversity decreases with latitude, which implies, if plant cover does not change significantly, the abundance per unit area/range of the individual species must increase with latitude. As most aphids are host specific and cannot survive off their host plants for very long Dixon *et al.* (1987) argue that their host plants have to be relatively abundant (p. 65). Because of the great diversity of plants in the tropics there are relatively few species, compared to the temperate regions, which are abundant enough to sustain specific aphids. As a consequence the diversity of aphids is greater in temperate than tropical regions. Therefore, aphid diversity and leaf colour may be independently correlated.

Alternative explanation

The most likely reason for autumnal leaf colour is that it functions as a sun screen. When plants are exposed to levels of irradiance in excess of that which they can utilize in photosynthesis, the absorption excess excitation energy can bring about photoinhibition. This manifests itself as a decrease in a plant's photosynthetic capacity. This is especially true when plants are exposed to other stresses, like low autumn temperatures, which may cause photoinhibition to occur even at relatively low irradiances (Wilkinson et al., 2002).

MUTUALISM BETWEEN APHIDS AND TREES

Aphids produce large quantities of honeydew, which in some cases contains a high percentage of the trisaccharide sugar melezitose (Bacon & Dickinson, 1957). A large proportion of this honeydew reaches ground level, which can result in as much as 10 g of sugar per 100 g of soil. This led Owen and Wiegert, in a series of papers (Owen, 1978, 1980a; Owen & Wiegert, 1976, 1981), to propose that trees release surplus sugars by enlisting the help of aphids. This sugar is used by free-living nitrogen-fixing bacteria in the soil, which increase in number beneath aphid infested trees and make more nitrogen available to these trees. Melezitose, or a particular mixture of sugars in honeydew, are thought to have an optimal effect on nitrogen fixation. The aphids are seen as a 'necessary part' of a tree, releasing surplus sugar that promotes a better supply of nitrogen.

The addition to the soil of the four sugars – fructose, glucose, melezitose and sucrose – commonly found in honeydew, at rates equivalent to those recorded beneath lime trees, causes an increase in the abundance of bacteria in woodland soils (Dighton, 1978a,b). In the laboratory, fructose is more effective at promoting nitrogen fixation than melezitose (Petelle, 1980). However, as Petelle used single sugars rather than a mixture his results do not refute Owen's hypothesis (Owen, 1980b). A more rigorous test of the mutualism hypothesis, in which alder aphids Pterocallis alni were removed from red alder Alnus rubra by spraying with malathion, revealed that an aphid infestation resulted in a decrease in ammonification and nitrification in the soil and a decrease in above-ground primary production (Grier & Vogt, 1990). Thus, contrary to the prediction of the hypothesis, nitrogen availability in the soil is apparently markedly reduced by large quantities of aphid honeydew and there is no positive effect on tree growth.

Much of the honeydew excreted by aphids feeding on the leaves and needles of trees falls on to the upper surface of other leaves where it promotes the growth of sooty moulds. In some years these sooty moulds blacken the upper surface of leaves. On pecan *Carya illinoensis* these moulds can reduce light penetration and photosynthesis by 25–98%. In addition, the darkening of the leaf surface can result in an increase in leaf temperature of 4 °C (Smith & Tedders, 1980; Tedders & Smith, 1976; Wood *et al.*, 1988). Epiphytic micro-organisms, which includes the sooty moulds, are one of the most abundant groups of organisms. In areas where there is a lot of industrial pollution rich in nitrogen (Gundersen *et al.*, 1988; Schulze, 2000) the limiting resource for epiphytic micro-organisms is energy (Tsai *et al.*, 1997). The needles and leaves of aphid-infested trees in such areas show a dramatic increase in the abundance of micro-organisms (bacteria, yeasts and filamentous fungi) of two to three orders of magnitude (Stadler *et al.*, 1998). In addition to the changes in abundance there are also changes in species composition, more so on the leaves of beech than on oak, probably due to differences in the surface micro-morphology of the leaves (Stadler & Michalzik, 2004).

The effects micro-organisms have on the chemical content of rain as it percolates through the canopy – i.e. throughfall chemistry – which is important because it determines the input of nutrients and ions into forest soils are poorly understood. For example, in June when aphids are most abundant on spruce *Picea sitchensis*, the concentration of dissolved organic carbon (DOC) in throughfall collected beneath infested spruce trees is high and declines with the subsequent decline in aphid numbers. There is a very high correlation between DOC concentrations in throughfall and aphid abundance. However, the concentration of dissolved organic nitrogen (DON) in throughfall increases after the aphid population peaks and starts to decline in abundance: possibly a consequence of the peak in the abundance of epiphytic micro-organisms occurring after the peak in the production of honeydew. Concentrations of inorganic nitrogen are lower in throughfall collected beneath spruce heavily infested with aphids compared with uninfested trees, however, it becomes similar as the aphid numbers decline.

Field experiments show that following a high input of DOC from the canopy there is an increase in the DOC concentration in forest soil solutions, which is slightly delayed and longer lasting than the above ground aphid infestation. Similarly, there is an increase in the concentrations of DON and NO_3-N in the forest floor solution beneath aphid infested trees. Laboratory experiments, in which simulated

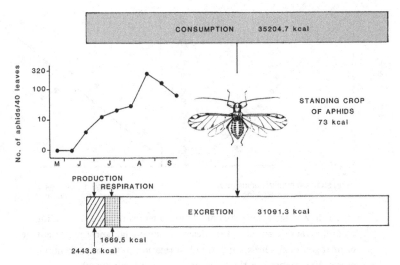

Figure 10.1. The annual consumption, excretion, respiration, production and standing crop of lime aphids in kilocalories on a 12 m lime tree. The inset graph is of the seasonal trend in aphid abundance used to calculate the energy flow. (After Llewellyn, 1972, 1975.)

artificial honeydew is applied to cores of forest soil, reveal that low to medium inputs of honeydew increase base respiration within one hour and cause a decline in NH_4-N fluxes. Large inputs of honeydew increase the immobilization of both NH_4-N and NO_3-N and slightly reduce DON fluxes. The DOC fluxes increase considerably but decline to the base level within 72 hours of applying honeydew. Thus, inorganic carbon from aphids in throughfall affects the mineralization, mobilization and transport of organic matter in forest soils (Stadler & Michalzik, 2004). That is, changes in aphid abundance can have a dramatic effect on the abundance of microbes and nutrient fluxes in forests.

ENERGY DRAIN IMPOSED BY APHIDS AND TREE GROWTH

As phloem sap contains high concentrations of sugar and very little amino-nitrogen aphids have to process very large quantities of sap in order to obtain sufficient amino-nitrogen to sustain their very high rates of growth. In the case of the giant willow aphid *Tuberolachnus salignus* Gmelin, a single aphid consumes the photosynthetic product of 5–20 cm^2 of leaf per day (Mittler, 1957). The annual drain imposed on a 14 m tall lime tree by a natural population of the lime aphid *Eucallipterus tiliae* L., is considerable (Figure 10.1). During the course of a year, the

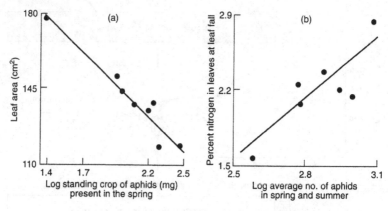

Figure 10.2. The average area of sycamore leaves relative to standing crop of sycamore aphids on the leaves in spring (a); and the percentage of nitrogen in the leaves at leaf fall in relation to the average number of sycamore aphids on the leaves in spring and summer (b).

population turns over its own standing crop 482 times, or 3.4 times per day (Llewellyn, 1970). This is considerably greater than that achieved by oribatid soil mites (38 times per annum) and grasshopper populations (10 times per annum) (Macfadyen, 1964; Smalley, 1960). Thus, although the lime aphid is not particularly effective at utilizing its energy intake, it turns over energy at a massive rate, much of which falls to the ground as honeydew. The annual production of honeydew by the lime aphid is equivalent in energy terms to 0.8 of that locked up in the leaves at leaf fall. In the case of the sycamore it is on average equal to the energy in the leaves at leaf fall.

In the presence of the wood ant *Formica rufa* L. the energy drain imposed on sycamore can change dramatically. This ant preys on the sycamore aphid and tends another aphid, *Periphyllus testudinaceus* found on sycamore. The sycamore aphid removes approximately three times as much sap from trees in the absence than in the presence of ants, whereas the ant-tended species removes up to 50 times more sap from ant-foraged trees than from unforaged trees. However, the ant-tended aphid on average only removes one-fifth of that removed by the sycamore aphid each year (Warrington & Whittaker, 1985).

Sycamore saplings that are aphid infested clearly grow markedly less than uninfested saplings. Although aphids do not affect the number of leaves borne by lime, oak or sycamore, when heavily infested in spring the sycamore produces smaller leaves that contain more nitrogen (Figure 10.2). However, the leaf area equivalent to the energy removed by sycamore aphids only accounts for a small proportion of

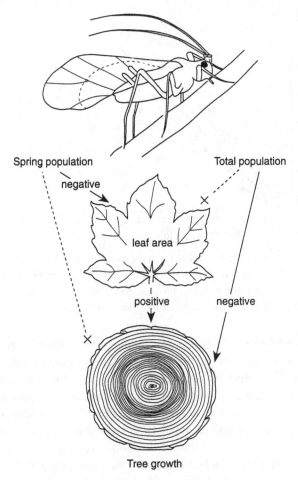

Spring population   Total population

negative

leaf area

positive   negative

Tree growth

Figure 10.3. Diagrammatic representation of the effect of sycamore
aphid numbers on leaf area and width of the annual rings of sycamore.

the observed diminution in leaf area. If the drain imposed is expressed
in terms of nitrogen rather than energy then aphids again remove far
less nitrogen than expected from the reduced size of the leaves (Dixon,
1971a). This implies that the effect aphids have on tree growth is not a
direct consequence of the energy and/or nutrient drain. Other factors,
for example the saliva aphids inject into plants may contain physiolog-
ically active components that adversely affect tree growth.

The width of the annual rings of sycamore is positively correlated
with the average size of the leaves, and negatively with the number of
aphids, on a tree throughout a year (Figure 10.3). This is possibly associated with the fact that each annual ring is composed of two types of

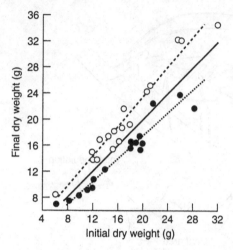

Figure 10.4. The effect of aphid infestation on the growth of oak *Quercus robur* saplings: their dry weight in autumn in relation to their estimated dry weight the previous spring for saplings infested (●) and kept free of aphids (○). The solid line indicates no change in weight.

vessel, which make up the spring and summer wood. The springwood is mainly laid down when the leaves are developing in spring, whereas the summerwood, which makes up most of each annual ring, is mainly laid down after the leaves stop growing. In the absence of aphids, some sycamore trees could produce as much as 280% more stem wood (Dixon, 1971a). Lime and oak aphids hatch later than the sycamore aphid, relative to the time of bud burst of their host trees, and as a consequence rarely become abundant before the leaves are fully grown. This is reflected in the fact that the aphids on these trees do not affect the above-ground growth in girth and stem length of their respective hosts. However, infested saplings of lime and oak *Quercus robur* often weigh less at the end of the year than they did at the start, mainly due to a reduction in the mass of their roots (Figure 10.4).

Aphid infestation causes early leaf fall in all three species, and in *Q. robur* and sycamore results in the leaves becoming a darker green. In oak this is a consequence of a 25% increase in the quantity of both chlorophyll A and B. Associated with this is an increase in dry matter production per unit area of leaf, which in sycamore can be 1.7 times greater in infested than in uninfested saplings. Following years of heavy aphid infestations lime and sycamore break their buds later than usual, and in the case of lime the leaves are smaller and a darker green, and have a net production 1.6 times greater than the leaves of previously uninfested saplings (Dixon, 1971b). Similar negative affects of aphids on the growth of other trees are reported in the literature (Table 10.1).

Although saplings and mature trees may differ in their reaction to aphid infestation, there is no doubt that aphids have a pronounced adverse effect on the growth of mature trees. Aphids could be a major factor determining the outcome of intra- and interspecific competition between trees, and the establishment of seedlings, as they often become heavily infested with aphids that fall from the parent trees. Some trees can compensate in part at least for the nutrient drain imposed by aphids and it remains to be shown whether aphids affect the fitness of mature deciduous trees. However, high numbers of the spruce aphid *Elatobium abientinum* can cause the defoliation of Sitka spruce, which, unlike Norway spruce, has only recently come into contact with the spruce aphid.

As well as imposing a severe nutrient drain, aphids transmit viruses, which could also adversely affect the fitness of trees. However, some genera and groups of woody plants are notable for not having aphid-transmitted viruses – e.g. Coniferae, Rosaceous trees and shrubs such as *Rosa, Malus* and *Pyrus*. During their long life trees must be exposed to millions of aphids and therefore subject to strong selection for tolerance or resistance to aphid-transmitted viruses. Depending on the type of virus – i.e. stylet borne or circulative – there are costs and benefits of virus transmission for aphids. In transmitting viruses to non-host plants growing in the immediate vicinity of the specific host plants, it has been suggested that aphids could indirectly improve the competitive status of their host plant (Eastop, 1991). However, the mechanism by which this is likely to have evolved is unclear.

APHIDS AND SEED PRODUCTION

The number and quality of seed produced by plants is used as a measure of genetic fitness. The fact that many trees do not fruit or seed equally every year, but at irregular intervals, makes any study of the effect of aphids on seed production difficult. However, there are reports in the literature of aphids negatively affecting both the yield and size of seed produced by trees (Table 10.1). In the case of sycamore, seed production over a period of eight years is not correlated with either aphid numbers in current or previous years, but is strikingly dominated by the very infrequent good years when seed production can be four times the long-term average. That is, there appears to be no evidence from

Table 10.1. Summary of the effect of other species of aphids on tree growth

			Effect of aphid on:			Growth		Seed	
Tree	Reference	Aphid	Photosynthesis	Chlorophyll	Time to leaf fall	Shoots	Roots	Yield	Size
Apple									
Malus sp.	5, 6, 14, 15	Aphis spiraecola Patch	-ve	-ve		-ve	-ve		
		Dysaphis plantaginea (Pass.)	-ve			-ve	-ve		
Maple									
Acer amoenum	2, 3	Periphyllus californiensis Shinji	-ve			-ve		-ve	
Pecan									
Carya illinoensis (Wang)	9–13, 16–18	Monellia caryella (Fitch)	-ve	-ve		-ve	-ve	-ve	-ve
		Melanocallis caryaefoliae (Davis)	-ve	-ve	-ve	-ve	-ve	-ve	-ve
		Monelliopsis pecanis (Bissell)	-ve	-ve		-ve	-ve	-ve	
Walnut									
Juglans regia L.	7	Chromaphis juglandicola (Kaltenbach)						-ve	-ve

Fir

Abies sachalinensis
Fr. Schmidt | 19 | Cinara todocola (Inouye) | -ve

Pine

Pinus taeda L | 1 | Cinara atlantica (Wilson) | -ve
| | Cinara watsoni Tissot | -ve

Spruce

Picea glehnii Mast | 4 | Cinara pruinosa ezonana Inouye | -ve

Picea sitchensis (Bong) Carr. | 8 | Elatobium abietinum (Walker) |

References: 1, Fox & Griffith (1977); 2 & 3, Furuta (1990b, 1994); 4, Furuta et al. (1983); 5 & 6, Kaakeh et al. (1992a,b); 7, Sibbett et al. (1982); 8, Straw et al. (1998); 9, Tedders (1978); 10–12, Tedders et al. (1981a,b); 13, Tedders & Wood (1985); 14, Varn & Pfeiffer (1989); 15, Wilkaniec (1990); 16, Wood & Tedders (1986); 17 & 18, Wood et al. (1985, 1987); 19, Yamaguchi & Takai (1977).

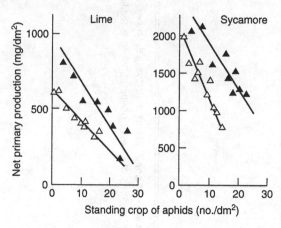

Figure 10.5. The net annual primary production per unit area of leaf in relation to the average number of aphids on the leaves during the year (standing crop) for lime and sycamore saplings watered with either full (▲) or half-strength (△) nutrient solutions. (After White, 1970.)

this study to support the suggestion of Leather (2000) that in years following heavy aphid infestations sycamore flowers are protandrous and less likely to bear seed (Binggelli, 1992) and largely escape infestation by aphids. Heavy aphid infestations at the time of flowering, however, can cause flowers to abort, which results in the production of relatively few but large seeds. That is, by reducing the seed crop in such years could have a positive effect because seed quality is greatly increased. Another complicating factor is the quality of the soil in which the trees are growing. Experiments with saplings indicate, nevertheless, that variation in aphid abundance is likely to have a more marked adverse effect on tree growth, and therefore possibly on seeding, than is soil quality (Figure 10.5).

RECRUITMENT

In another maple, *Acer saccharum*, which also produces a variable seed crop, there is a strong density-dependent relationship between the proportion of seeds surviving and the number shed (Harper, 1977). Sycamore seeds germinate early and the seedlings are common beneath and around sycamore trees, occasionally forming dense carpets in woodlands. However, as soon as the buds of the trees burst and the canopy closes most of these seedlings die. This appears to be mainly a consequence of their being in the shade. As in *A. saccharum*, recruitment in sycamore is likely to be dependent more on where seedlings happen by chance to be located on the woodland floor than on the number of seedlings. Therefore, it is unlikely that aphids affect the number of trees recruited. It is more likely, however, that aphids through their effect on the growth of saplings determine which will survive intra- and interspecific competition and become mature trees. This conforms with the view expressed by Crawley (1983), who looked at the role of herbivores, in general, in regulating plant abundance.

In summary, although the idea that deciduous trees signal their level of commitment to defence against aphids by the colour of their leaves in autumn is both attractive and popular, there are several facts relating to aphid ecology that are at odds with this theory. Similarly, there is no evidence to support the proposed mutualism between aphids and trees, which supposedly increase the availability of nitrogen for tree growth. However, honeydew produced by aphids does affect the epiphytic flora growing on the leaves and needles of trees and the nutrient flow through forest ecosystems, which can result in the immobililzation of nitrogen in soil rather than increasing its availability. Although the considerable energy drain imposed by aphids has a very marked negative effect on tree growth and seeding, it is unlikely to affect sycamore recruitment. However, their negative effect on tree growth is likely to be important in determining which saplings survive competition and develop into mature trees.

11

Rarity, conservation and global warming

Common species tend to be taken for granted and excite little interest, except when they become extremely abundant. Rare species, however, tend to attract more attention especially if they are large and/or colourful. Insects rarely fall into this category, and however rare are unlikely to become the subject of a conservation programme. That is, for most people, for example, one species of aphid is much like another and many are pests so why should one want to conserve them. Hopefully, insects will eventually be regarded as beautiful and fascinating, and worth conserving in their own right (Holloway, 2003), but this is unlikely except for a few large and/or colourful species. Meantime, however, insect conservation is likely to have to depend on the conservation of 'flagship species' or 'umbrella species', which are usually mammals, to safeguard certain habitats that incidentally include many other species of organisms, especially insects.

RARITY

Although rare insects are unlikely to excite the interest of conservationists they could be good subjects for studying rarity. Early in the twentieth century Garthside (1928) stated 'one of the outstanding facts is the extremely large number of species which occur in very small numbers'. This sort of observation attracted the attention of biomathematicians like Williams (1964) who thought that a mathematical model that closely fits the observed data would increase our theoretical understanding of rarity and result in further tests and experiments. He then went on to show that the frequency distribution of species with different numbers of individuals is of the 'hollow curve' type. That is, a few species are very abundant but the majority are uncommon to rare.

Table 11.1. *A classification of species abundance based on three characteristics, i.e. geographical range, feeding specificity and local population size with examples of the five classes of rarity applicable to aphids*

Feeding specificity	Generalist	Specialist	Generalist	Specialist
Geographical range	Large	Large	Small	Small
Population size				
Locally abundant	*Aphis fabae*	*Euceraphis betulae*	a	*Aneocia corni*
Constantly sparse	a	*Monaphis antennata*	a	*Nasonovia saxifragae*

(After Rabinowitz, 1981.)
a, Not applicable to aphids.

As most insects are specialists and associated with particular habitats, their rarity could be a consequence of the rarity of the habitat. The theoretical basis for this is presented in Chapter 6 (pp. 64–5). This raises the question of the meaning of rarity and whether relative to their resources any species is truly rare. Gaston's (1994) definition of rarity as those species that are in the lower 25% quartile of the abundance distribution, however, does little to further our understanding of why some organisms are rare. That is, there is a need to define the 'niche space' occupied by each species. As most aphids tend to live on one or at most a few species of the same genus of host plant, their habitat (p. 18) and relationships with their host plant and natural enemies – their niche space – is well defined (p. 43).

When asking why particular species are rare it has proved useful to consider rarity at three levels: geographical range, host specificity and local population size. This gives eight classes (Table 11.1). The only truly common species of aphids are the few generalist aphids, like *Aphis fabae*, which have a large geographical range and are locally abundant in several habitats. The specialist species can be assigned to one of four remaining seven forms of rarity given in Table 11.1. Several of the classes of rarity involving specialist aphids can be accounted for in terms of the abundance of their host plants (pp. 64–5). Host-specific aphids that live on uncommon plants are likely to incur great losses in finding their host plants and as a consequence are rarer than aphids living on common plants. However, some species have a wide geographical range, live on a very abundant host but always occur at a very low density. For example, *Monaphis antennata* (Kaltenbach) feeds on birch *Betula* spp. and occurs throughout the Palearctic, from Finland to Portugal, the UK to Japan, and yet is rare everywhere. This aphid is an enigma, as by definition it deviates from the general trend for range and abundance to be positively correlated (Gaston & Lawton, 1990) and it does not appear to have a 'core' to its range where it is particularly common (Hengeveld & Hacek, 1982).

Hopkins (1996) made a detailed study of *Monaphis* and the other species of aphids living on birch in an attempt to account for the rarity of this aphid. Of the species of aphids that live on birch in the UK only three are frequently caught by suction traps of the Rothamsted Insect Survey. They are *Euceraphis* spp., *Betulaphis* spp. and *Kallistaphis flava*. Their relative aerial abundance at each of the suction trap sites throughout the UK appears to be similar. Too few individuals of any of the other birch aphids were caught to allow meaningful comparisons, and *Monaphis* was the only species not caught by these traps. This tends to confirm that although present in England, Wales and Scotland, *Monaphis* is less abundant than the other birch aphids.

In comparison with six other species of aphids that live on birch, *Monaphis* differs in that it feeds on the upper surface of leaves (Hopkins & Dixon, 1997), is nearly semelparous and produces relatively small offspring. The rapid birth of a single batch of offspring (semelparous) rather than of individual offspring at intervals over a long period of time (iteroparous) is rare in aphids. Semelparous reproduction in aphids is associated with either a constraint on time or a high incidence of adult mortality (Dixon & Dharma, 1980; Walters *et al.*, 1984). The intrinsic rate of population increase of *Monaphis*, however, is only slightly lower than that of the most common species, *Euceraphis*, on birch. No feature of its environment, including natural enemies, appears to limit its abundance and it is able to live in a broad range of conditions, on several species of birch of different ages.

At least three reasons could be proposed for why *Monaphis* is different. The differences may be due to phylogenetic history, irrespective of selection pressures. They may reflect an ecological strategy: rarity. Or, external factors may account for the rarity, and the observed differences are a consequence of the rarity.

Some models of evolution assume that characters change in a random fashion. If this is the case then the magnitude of changes from the ancestral state is proportional to the time since it diverged from other species (Martins, 1996). That *Monaphis* shows the greatest character change in the Drepanosiphinae but is of similar evolutionary age or younger than most other species argues for these unusual traits being the product of directional selection.

Rarity as a strategy would appear to require group selection, which is regarded as 'unnecessary and impossible' (Gilpin, 1975) and not a valid model of selection (Maynard-Smith, 1976; Krebs & Davies, 1987). As far as *Monaphis* is concerned there would appear to be no force acting to prevent or penalize any individual from cheating and developing 'normal' characters and increasing in abundance.

Rabinowitz *et al.* (1984) argue that the differences between rare and common species are not the cause of rarity but due to an external factor and are therefore adaptive. The unusual traits of *Monaphis* could be viewed as a response to an extinct parasitoid (historical) or an extant, as yet unrecorded parasitoid (contemporary). They could have evolved as non-adaptive consequences of the rarity due to an external factor, as an adaptive response to an external factor, or as an adaptive response to the state of rarity.

However, none of the above is convincing given the available evidence. It is possible that *Monaphis* is on a trajectory to extinction, with the apparent absence of any other aphid with these traits being because they are already extinct. It is difficult to reconcile *Monaphis'* supposedly pending ecological failure (i.e. extinction) with its ecological success in terms of its extensive geographical range. There is some evidence that traits that influence commonness and rarity are conserved at taxonomic levels above species (Hodgson, 1986; Nee *et al.*, 1991; Ricklefs & Latham, 1992) and that many ecological traits are relatively static (Coope, 1995). However, why should *Monaphis* be the only species of aphid to respond to selection in this way?

If species living in the same habitat share natural enemies then they may have a more marked effect on the abundance of the uncommon than of the common species, a phenomenon known as apparent competition. The potential importance of apparent competition in community structure was discussed by Holt and Lawton (1994) and it is thought that communities of herbivorous insects might be structured by apparent competition mediated by shared predators and parasitoids (Jefferies & Lawton, 1984; Lawton, 1986; Godfray, 1994). The poorer performance of nettle aphid colonies adjacent to heavily aphid infested grass plots is attributed to predation by high numbers of ladybirds attracted to the area by the grass aphids (Müller & Godfray, 1997). However, *Monaphis* relative to the other species on birch appears to occupy 'enemy-free' space (Hopkins & Dixon, 1997) and even if apparent competition is important it is unlikely to keep a species rare indefinitely.

Monaphis is an unusual aphid with specific life-history traits and niche preferences, which may be associated with predator pressure. However, data for other rare aphids are needed to establish whether life history is the cause or just reflects the selection pressures experienced by rare species. Although there are unequivocal examples of niche restriction causing rarity, what is known about *Monaphis* argues that this need not necessarily be the case for all rare species. A major stumbling block to interpreting the results is resolving whether the observed differences are the cause or consequence of rarity – this must be addressed if a general theory of rarity is to be constructed.

CONSERVATION

Whether or not a species of plant hosts aphids is thought to be determined mainly by its abundance or the proportion of ground it covers (Dixon *et al.*, 1987). It is likely that the more abundant a plant the more species of aphids it hosts, with an abundant host like birch hosting up to eight species of aphids in the UK. Support for this comes from an analysis of the proportion of the aphid fauna living on particular species of trees in Europe that are present in the UK relative to the range of the trees in the UK (pp. 167–8). Host range and abundance are usually correlated. The greater the range of the host tree in the UK the greater the proportion of the species of aphid that are present (Figure 11.1). That is, the hosts of host-specific European aphids that do not occur in the UK are rarer there. This relationship, however, may also be partly determined by those trees with a small range in the UK being at the edge of their range and growing in conditions that are unfavourable for some of the aphids that live on it in Europe.

The previous discussion tends to indicate that it is highly likely that many phytophagous insects are rare because their host plants are rare. This idea was tested by using the information available on the rarity of species in the UK of four well-studied insect taxa (macromoths, gelechiid micro-moths, beetles and tephritid flies), some species of which have been afforded conservation status (Shirt, 1987). The proportion of species that are rare in these taxa declines as host plant range increases. By extrapolating these patterns of host range/ phytophagous rarity to aphids it is possible to identify species that are probable candidates for conservation.

The first step in extrapolating rarity from well-known taxa to poorly known taxa is to establish the relationship between insect

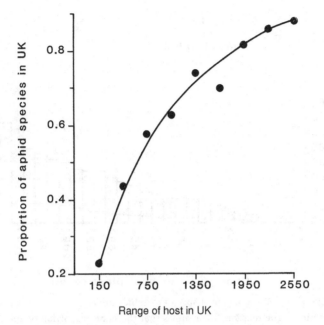

Figure 11.1. The proportion of the aphid fauna present on nine species of trees in Europe that are present on those trees in the UK relative to the ranges of those trees in the UK.

abundance and host range for the well-known taxa. Across categories of increasing host range, the proportion of species that are rare is expected to decline from a point where all species are rare on very rare hosts to the overall proportion of rarity for the taxon. Identifying the species that are rare in a poorly known taxon requires that a threshold of host range is determined below which a certain proportion of the well-known taxon is rare. When this host range threshold is applied to the poorly known taxon the same proportion of species whose hosts are below this limit can be presumed to be rare.

The choice of well-known taxa to predict patterns of rarity in lesser-known taxa depends on their biological and ecological similarity. Among the ecological characteristics of insects that are thought to be most important in determining their ability to use rare hosts is their abilty to find them (Dixon et al., 1987), which is influenced by flight strength and ability to survive away from hosts. Using the pattern of host-use by species in the well-studied taxa, it is possible to identify species of likely conservation concern in these taxa. This is achieved by

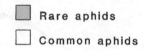

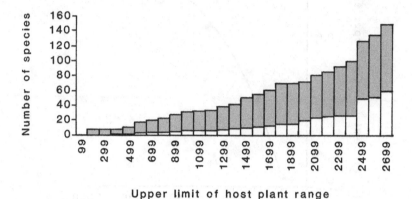

Upper limit of host plant range

Figure 11.2. The number of species of aphids predicted to be rare or common. The relationship assumes that the same proportion of aphid species are rare for a given host range as has been measured in the gelechiids.

identifying the range thresholds below which most of the species are of conservation concern in the well-studied taxa, and then identifying which species in the poorly studied taxon utilize hosts with similar ranges.

The proportion of insects that are of conservation concern is very high on the rarest hosts, but when progressively commoner hosts are examined the proportions of insects of conservation concern decreases as a smooth function (Figure 11.2). Therefore, it is not possible to identify a host-range threshold below which all species are rare, but it is possible to identify range thresholds below which most species are likely to be rare.

Host plant rarity is assessed by geographical range, measured as the number of 10-km squares in which this species has been recorded. There are 2862 10-km squares in the UK. For gelechiids 85% or more of the species are rare on host plants with a host plant range equal to or below 899 10-km squares. Gelechiid moths were chosen for this estimate because they are like aphids in being weak fliers. Whilst this is a subjective process, similar results are obtained if other indicator groups are used, for example using the 85% level for gelechiids generates a list of 27 aphids, and the same number if beetle data are used. Empirical data indicate that most of these aphids are indeed rare. Therefore, this method provides an economical and objective way of producing a shortlist of aphids rare enough to warrant conservation (Hopkins *et al.*, 2002). Of the 27 aphids only one *Myzocallis carpini* (Koch) lives on a tree – hornbeam *Carpinus betulinus*.

CO-EXTINCTION

One way to estimate the extinction threat faced by poorly known species is to identify whether they possess ecological traits that are likely to make them particularly vulnerable to extinction, as for example when one species is dependent on a second species, which is itself rare or threatened. Extinction of the second species will lead to the extinction of the first, a phenomenon that Stork & Lyal (1993) termed co-extinction. The world's threatened flora is estimated to comprise 22–47% of plant species (Pitman & Jorgensen, 2002), so if threatened plants host aphids then co-extinction could be a substantial threat to aphid diversity.

Of the 1595 species of tree aphids, 11 (0.69%) are restricted to threatened trees. The 11 species of aphids live on nine species of trees (Table 11.2). One of these, *Quercus dumosa*, which hosts the aphid *Tuberculatus passalus*, is endangered and facing a very high risk of extinction in the wild in the near future. Four species of trees, with six species of aphids, are classed as vulnerable, meaning they face a high risk of extinction in the wild in the medium-term future. Four species of trees, with one specialist aphid species each, are classified as lower risk (near threatened), which means they are close to qualifying for the vulnerable classification.

Up to 47% of plant species (Pitman & Jorgensen, 2002) and 25% of insect species (McKinney, 1999) could be at risk of extinction. Thus even if the 1% of aphids threatened with co-extinction following tree extinction is extrapolated to herbivorous insects generally it is only a small percentage of those perceived to be threatened with extinction. There could be at least two reasons for this. First, some insect herbivores have very demanding habitat requirements and are rare even though their host is common. Second, even if plant extinction is the primary threat to an insect species, the extinction of the herbivore may occur long before the plant is considered rare enough to be included on the threatened species lists. In the UK, insect herbivore rarity is a continuous function of host plant rarity (Figure 11.2), but it is not clear how rare

Table 11.2. Aphids on threatened trees

Tree	Category of host	Host distribution	Source of threat	Aphids	Aphid distribution
Quercus dumosa	Endangered	Mexico, USA	Habitat loss	Tuberculatus passalus	California, USA
Picea morrisonicola	Vulnerable	Taiwan	Exploitation	Prociphilus formosanus	Taiwan
Pinus albicaulis	Vulnerable	USA, Canada	Habitat loss	Cinara anzai	California, Oregon, USA
		USA, Canada	Habitat loss	Cinara inscripta	Oregon, USA, British Columbia, Canada
		USA, Canada	Habitat loss	Cinara oregoni	Oregon, USA, British Columbia, Canada
Podocarpus salignus	Vulnerable	Chile	Exploitation and habitat loss	Neophyllaphis podocarpini	Chile
Lithocarpus indutus	Vulnerable	Indonesia	Few, small populations	Sinonipponaphis hispida	Java
Cercidiphyllum japonicum	Lower risk (near threatened)	China, Japan	Poor regeneration	Aulacorthum cercidiphylli	Japan
Cinnamomum japonicum	Lower risk (near threatened)	China, Korea, Japan	Habitat loss	Thoracaphis sp.	Japan
Keteleeria fortunei	Lower risk (near threatened)	China, Viet Nam	Not stated	Cinara keteleeriae	Yunnan, China
Zelkova carpinifolia	Lower risk (near threatened)	Azerbaijan, Georgia, Iran, Turkey	Not stated	Byrsocryptoides zelkovaecola	Georgia

a host must be before this is the most important threat to the insect (Thacker *et al.*, 2003).

Much of the debate about the effect climate change may have on insect abundance is concerned with its possible differential effect on bud-burst in plants and egg hatch in insect herbivores (Dewar & Watt, 1992; Harrington *et al.*, 1999; Watt & Woiwod, 1999). Watt & McFarlane (2002) in their recent review challenge the conclusion of Buse & Good (1996) that climate change has no effect on phenological synchrony. Buse & Good (1996) used saplings growing in pots and Watts & McFarlane (2002) argue that the roots of these saplings are likely to have experienced similar temperatures to the rest of the plant, which is unlikely to be the case for deep-rooted mature trees. Although an interesting point, it is worth asking: Is synchrony per se the issue we should be addressing? What is possibly more important is the mechanism(s) by which herbivores and hosts track the seasons and whether herbivores can quickly adapt to changes in the phenology of their host. That insects are well synchronized with their host plants indicates that adaptation is possible. Compared to many other insects aphids are a good model group for exploring this problem as the mechanisms by which they track the phenology of their host plants and their population dynamics are well understood.

There is a lot of variability in both the time of egg hatch and bud burst, both within and between years. Those eggs that hatch at the time of bud-burst tend to be fitter than those that hatch earlier or later (pp. 22–4; Dixon, 1976a). In addition, the time of egg hatch appears to be inherited (Komazaki, 1986; Mittler & Wipperfurth, 1988; Komatsu & Akimoto, 1995). Both times of bud-burst of sycamore saplings and egg hatch of the sycamore aphid are influenced by spring temperatures (Chambers, 1979). Both for trees and eggs there is an inverse relationship between the number of chill days experienced and the thermal time from some date in spring to bud burst and egg hatch (Mittler & Wipperfurth, 1988: Murray *et al.*, 1989). That is, both buds and eggs tend to remain dormant until they experience their chilling requirement, and then the time to bud burst or egg hatch is a function of the intensity of the chilling they have experienced and subsequent spring temperatures. Although the lower developmental thresholds of sycamore saplings (1 °C) and sycamore aphids (5 °C) in spring differ (Chambers, 1979), nevertheless, egg hatch and bud burst are synchronized (Dixon,

1976a). That is, although the tracking mechanisms used by the plant and insect differ in detail, synchronization is possible. Selection would appear to have resolved a complex optimization problem.

Harrington *et al.* (1999) in their review of this subject highlight what they regard as research priorities. One such priority is: 'How will changes in synchronization affect population dynamics?' This is important because it has implications for the evolution of many herbivore plant interactions. In the case of the sycamore aphid there is an inverse relationship between abundance in spring and autumn – the 'see saw' effect. Low numbers in spring are followed by high numbers in autumn and vice versa (pp. 70, 84–6; Dixon, 1970c; Dixon & Kindlmann, 1998b). If climate change differentially affects bud burst and egg hatch in this system then it is likely the mortality of the aphids hatching in spring would be higher and the average fitness of the survivors lower (Dixon, 1976a). Climate-change-induced asynchrony in the frequency distributions in time of bud-burst and egg hatch in any one year is likely to be slight and any negative effect on aphid numbers compensated for within the year due to the dynamics of the system. In addition, proportionally more of the survivors will be individuals with inherited responses that enable them to track more closely the phenology of their host plant. These aphids are likely to mate with one another and as a consequence, the following year, proportionally more of the eggs hatch in synchrony with bud burst. That is, selection will correct for any asynchrony between egg hatch and bud burst.

Predictions of models that attempt to determine the effect of climate change on the abundance of an insect, which do not include the mechanisms by which the insect's abundance is regulated and how selection is likely to shape its life-history strategy, are unlikely to be realistic. For tree-dwelling aphids, where the processes acting at both the individual and the population level are fairly well understood, the evidence tends to indicate that the adverse effect of climate change is unlikely to be as threatening as is so often painted (Dixon, 2003).

In summary, rarity in aphids appears to be mainly a consequence of living on rare host plants. However, some, like *Monaphis antennata*, live on widely distributed and abundant host plants but nevertheless are rare everywhere. This aphid has unique life-history traits and niche preferences. Until, however, other similarly rare species are studied in detail it is not possible to say whether these differences are the cause or a consequence of rarity. Using information on well-studied insect taxa, which have a similar ecology to aphids, it is possible to show that the proportion of species that are rare in these taxa declines as host-plant range increases. Extrapolating this relationship between insect abundance and host range to aphids indicates that 27 aphid species in the UK are likely to be rare because their host plants are rare. This is supported by empirical data, and one of the species is a tree-dwelling aphid. Of the 1600 species of tree-dwelling aphids in the world, 11 are restricted to species of trees that are threatened with extinction. Of these, one, *Tuberculatus passalus*, is facing a high risk of extinction in the wild in the near future. Global warming, through its differential effect on bud-burst in plants and egg hatch in insect herbivores, is perceived as a threat to the abundance of tree-dwelling insects in particular. However, it is likely that in aphids selection will quickly correct for any asynchrony between egg hatch and bud burst.

Epilogue

Studies on insect herbivore–host dynamics have mainly been on leaf-chewing insects and their host plants. This can be justified because many of these insects are relatively large and frequently defoliate, and manifestly adversely affect the fitness of, their host plants. The sap-sucking insects, by contrast, tend to be inconspicuous and apparently less damaging. However, aphids in particular are often very abundant and the most important component of the herbivore fauna of the canopy of trees in boreal and temperate forests, 'one of the least explored zones on land'. For this reason alone the study of tree-dwelling aphids is justified.

Has this study revealed any broad generalization or principle? It lends strong support to the generation time ratio (GTR) concept, which proposes that the most important life-history trait determining the outcome of the interaction between an insect herbivore and its natural enemies is their relative developmental times. Only those natural enemies that have developmental times similar to a hebivore are likely to regulate its abundance. Surprisingly, even the parasitoids in the systems studied here have appreciably longer developmental times than the aphids. Even for pathogens this concept could apply if the interaction is viewed in terms of the GTR of the pathogen and that of the response time of the immune system of the host. As the GTR concept appears to operate in a wide range of insect predator/prey, parasitoid/host and pathogen/host systems it could qualify as a principle. In addition, self-regulation through competition for resources and a within-season varying carrying capacity are likely to be general features of aphid population dynamics, rather than either top–down or bottom–up regulation.

When abundant, aphids can impose a very severe nutrient drain on trees, which results in reduced growth and possibly poor seed production. Although the effect on growth could affect the competitive status of young trees, it appears unlikely that it greatly affects the competitive status or fitness of mature trees. That is, the interaction between aphids and trees is asymmetrical, with trees greatly affecting the life-history strategies and population dynamics of aphids but the latter having very little effect on trees.

In terms of empirical data it is clear there is a shortage of long-term, and in particular, extensive population censuses. Experience indicates that this information should not be seen mainly as a means of testing theory. The data collected should also include detailed information on a wide range of aspects of the population biology of the component species, as the better one understands the system the better the chance of developing a more appropriate theory. Using mathematics to describe the dynamics of the system brings rigour to the procedure. However, one should not be seduced by the elegance of mathematical models, but be more sceptical of what they have contributed to understanding nature.

References

Adams, D. & Douglas, A. E. (1997). How symbiotic bacteria influence plant utilization by the polyphagous aphid, *Aphis fabae*. *Oecologia* **110**, 528–532.

Alverson, D. R. & English, W. R. (1990). Dynamics of pecan aphids, *Monelliopsis pecanis* and *Monellia caryella*, on field-isolated single leaves of pecan. *Journal of Agricultural Entomology* **7**, 29–38.

Andrewartha, H. G. & Birch, L. C. (1954). *The Distribution and Abundance of Animals*. Chicago: University of Chicago Press.

(1984). *The Web of Life*. Chicago: University of Chicago Press.

Archetti, M. (2000). The origin of autumn colours. *Journal of Theoretical Biology* **205**, 625–630.

Ayal, Y. & Green, R. F. (1993). Optimal egg distribution among host patches for parasitoids subject to attack by hyperparasitoids. *American Naturalist* **141**, 120–138.

Bacon, J. S. D. & Dickinson, B. (1957). The origin of melezitose: a biochemical relationship between the lime tree (*Tilia* spp.) and an aphid (*Eucallipterus tiliae* L.). *Biochemical Journal* **66**, 289–299.

Barlow, N. D. (1977). A simulation study of lime aphid populations. PhD Thesis, University of East Anglia.

Barlow, N. D. & Dixon, A. F. G. (1980). *Simulation of Lime Aphid Population Dynamics*. Wageningen: Pudoc.

Begon, M. & Mortimor, M. (1981). *Population Ecology: A Unified Study of Animals and Plants*. Oxford: Blackwell Scientific Publications.

Behrendt, K. (1963). Über die Eidiapause von *Aphis fabae* Scop. (Homoptera, Aphidiae). *Zoologische Jahrbücher Physiologie* **70**, 309–398.

Berryman, A. A. (1981). *Population Systems: A General Introduction*. New York: Plenum.

Berryman, A. A. (2002). Population: a central concept for ecology? *Oikos* **97**, 439–442.

Binggelli, P. (1992). Patterns of invasion of sycamore (*Acer pseudoplatanus L.*) in relation to species and ecosystem attributes. PhD Thesis, University of Ulster.

Bonner, J. T. (1988). *The Evolution of Complexity by means of Natural Selection*. Princeton: Princeton University Press.

Bonnet, C. (1745). *Traité d'Insectologie au observations sur Pucerons*. Paris: Chez Durand.

Bronowski, J. (1973). *The Ascent of Man*. London: British Broadcasting Corporation.

Brown, M. (1975). Intra-specific mechanisms regulating the numbers of lime aphid. PhD Thesis, University of Glasgow.

Buchner, P. (1955). Endosymbiosestudien an Schildlausen 11 *Stictococcus diversiset*. *Zeitschrift für Morphologie und Okologie der Tiere* **43**, 397–424.

(1965). *Endosymbiosis of Animals with Plant Microorganisms*. New York: Interscience Publishers.

Bumroongsook, S. & Harris, M. K. (1991). Nature of conditioning effect on pecan by the blackmargined aphid. *Southwestern Entomologist* **16**, 267–275.

(1992). Distribution, conditioning, and interspecific effects of blackmargined aphids and yellow pecan aphids (Homoptera: Aphidiae) on pecan. *Journal of Economic Entomology* **85**, 187–191.

Buse, A. & Good, J. E. G. (1996). Synchronization of larval emergence in winter moth (*Operophtera brumata* L.) and budburst in pedunculate oak (*Quercus robur* L.) under simulated climatic change. *Ecological Entomology* **21**, 335–343.

Chambers, R. J. (1979). Simulation modelling of a sycamore aphid population. PhD Thesis, University of East Anglia.

Chang, K.-G., Fechner, G. H. & Schroeder, H. A. (1989). Anthocyanins in autumn leaves of quaking aspen in Colorado. *Forest Science* **35**, 229–236.

Charnov, E. (1982). *The Theory of Sex Allocation*. Princeton: Princeton University Press.

Charnov, E. & Bull, J. J. (1989a). Non-Fisherian sex ratios with sex change and environmental sex determination. *Nature* **338**, 148–150.

(1989b). The primary sex ratio under environmental sex determination. *Journal of Theoretical Biology* **139**, 431–436.

Cherrett, J. M. (1988). Ecological concepts: a survey of the members of the British Ecological Society. *Biologist* **35**, 64–66.

Collins, M. D. (1981). Coexistence in aphid parasites. PhD Thesis, University of East Anglia.

Coope, G. R. (1995). Insect faunas in ice age environments: why so little extinction? In *Extinction Rates*, ed. J. H. Lawton & R. M. May. Oxford: Oxford University Press, pp. 55–74.

Crawley, M. J. (1983). *Herbivory*. Oxford: Blackwell Scientific Publications.

Curtis, J. (1845). Observations on the natural history and economy of various insects affecting corn-crops, including the parasitic enemies of the wheat-midge, the thrips, wheat-louse, wheat-bug and also the little worm called *Vibrio*. *Journal of the Royal Agricultural Society of England* **6**, 493–518.

Daag, J. L. (2002). Strategies of sexual reproduction in aphids. PhD Thesis, University of Göttingen.

Dadd, R. H. & Krieger, D. L. (1968). Dietary amino acid requirements of the aphid *Myzus persicae*. *Journal of Insect Physiology* **14**, 741–764.

Dahlsten, D. L., Zuparko, R. L., Hajek, A. E., Rowney, D. L. & Dreistadt, S. H. (1999). Long-term sampling of *Eucallipterus tiliae* (Homoptera: Drepanosiphidae) and associated natural enemies in a northern California site. *Environmental Entomology* **28**, 845–850.

Day, K. & Crute, S. (1990). The abundance of spruce aphids under the influence of an oceanic climate. In *Population Dynamics of Forest Insects*, ed. A. D. Watt, S. R. Leather, M. D. Hunter & N. A. C. Kidd. Andover: Intercept, pp. 25–33.

de Equileor, M., Grimaldi, A., Tettamanti, G., Valvassori, R., Leonardi, M. G., Giordana, B., Tremblay, E., Digilio, M. C. & Pennachio, F. (2001). Larval anatomy and structure of absorbing epithelia in the aphid parasitoid *Aphidius ervi* Haliday (Hymenoptera: Braconidae). *Arthropod Structure and Development* **30**, 27–37.

Dempster, J. P. (1975). *Animal Population Ecology*. London: Academic Press.

Dewar, R. C. & Watt, A. D. (1992). Predicted changes in the synchrony of larval emergence and budburst under climatic warming. *Oecologia* **89**, 557–559.

Dighton, J. (1978a). Effects of synthetic lime aphid honeydew on populations of soil organisms. *Soil Biology and Biochemistry* **10**, 369–376.

(1978b). *In vitro* experiments simulating the possible fates of aphid honeydew sugars in soil. *Soil Biology and Biochemistry* **10**, 53–57.

Digilio, M. C., Isidora, N., Tremblay, E. & Pennacchio, F. (2000). Host castration by *Aphidius ervi* venom proteins. *Journal of Insect Physiology* **46**, 1041–1050.

Dixon, A. F. G. (1958). The escape responses shown by certain aphids in the presence of the coccinellid *Adalia decempunctata* (L.). *Transaction of the Royal Entomological Society London* **110**, 319–334.

(1963). Reproductive activity of the sycamore aphid, *Drepanisiphum platanoides* (Schr.) (Hemiptera, Aphididae). *Journal of Animal Ecology* **32**, 33–48.

(1966). The effect of population density and nutritive status of the host on the summer reproductive activity of the sycamore aphid, *Drepanosiphum platanoides* (Schr.). *Journal of Animal Ecology* **35**, 105–112.

(1969). Population dynamics of the sycamore aphid *Drepanosiphum platanoides* (Schr.) (Hemiptera: Aphididae): migratory and trivial flight. *Journal of Animal Ecology* **38**, 585–606.

(1970a). Factors limiting the effectiveness of the coccinellid beetle, *Adalia bipunctata* (L.), as a predator of the sycamore aphid, *Drepanosiphum platanoides* (Schr.). *Journal of Animal Ecology* **39**, 739–751.

(1970b). Quality and availability of food for a sycamore aphid population. In *Animal Populations in Relation to their Food Resources*, ed. A. Watson. Oxford: Blackwell, pp. 271–287.

(1970c). Stabilization of aphid populations by an aphid induced plant factor. *Nature* **227**, 1368–1369.

(1971a). The role of aphids in wood formation. I. The effect of the sycamore aphid, *Drepanosiphum platanoides* (Schr.) (Aphididae), on the growth of sycamore, *Acer pseudoplatanus* (L.). *Journal of Applied Ecology* **8**, 165–179.

(1971b). The role of aphids in wood formation. II. The effect of the lime aphid, *Eucallipterus tiliae* L. (Aphididae), on the growth of lime, *Tilia x vulgaris* Hayne. *Journal of Applied Ecology* **8**, 393–399.

(1971c). The life-cycle and host preferences of the bird cherry-oat aphid, *Rhopalosiphum padi* L., and their bearing on the theories of host alternation in aphids. *Annals of Applied Biology* **68**, 135–143.

(1971d). The 'interval timer' and photoperiod in the determination of parthenogenetic and sexual morphs in the aphid, *Drepanosiphum platanoides*. *Journal of Insect Physiology* **17**, 251–260.

(1971e). The role of intra-specific mechanisms and predation in regulating the numbers of the lime aphid, *Eucallipterus tiliae* L. *Oecologia* **8**, 179–193.

(1972a). The 'interval timer', photoperiod and temperature in the seasonal development of parthenogenetic and sexual morphs in the lime aphid, *Eucallipterus tiliae* L. *Oecologia* **9**, 301–310.

(1972b). Control and significance of the seasonal development of colour forms in the sycamore aphid, *Drepanosiphum platanoides* (Schr.). *Journal of Animal Ecology* **41**, 689–697.

(1974). Changes in the length of the appendages and the number of rhinaria in young clones of the sycamore aphid, *Drepanosiphum platanoides*. *Entomologia Experimentalis et Applicata* **17**, 1–8.

(1975a). Effect of population density and food quality on autumnal reproductive activity in the sycamore aphid, *Drepanosiphum platanoides* (Schr.). *Journal of Animal Ecology* **44**, 297–304.

(1975b). Seasonal changes in fat content, form, state of gonads and length of adult life in the sycamore aphid, *Drepanosiphum platanoides* (Schr.). *Transactions of the Royal Entomological Society of London* **127**, 87–99.

(1976a). Timing of egg hatch and viability of the sycamore aphid, *Drepanosiphum platanoides* (Schr.), at bud burst of sycamore, *Acer pseudoplatanus* L. *Journal of Animal Ecology* **45**, 593–603.

(1976b). Factors determining the distribution of sycamore aphids on leaves in summer. *Ecological Entomology* **1**, 275–278.

(1977). Aphid ecology: life cycles, polymorphism and population regulation. *Annual Review of Ecology and Systematics* **8**, 329–353.

(1979). Sycamore aphid numbers: the role of weather, host and aphid. In *Population Dynamics*, ed. R. M. Anderson, B. D. Turner & L. R. Taylor. Oxford: Blackwells, pp. 105–121.

(1985). Structure of aphid populations. *Annual Review of Entomology* **30**, 155–174.

(1987a). Cereal aphids as an applied problem. *Agricultural Zoology Reviews* **2**, 1–57.

(1987b). Adaptive significance of cyclical parthenogenesis in aphids. In *Aphids, their Biology, Natural Enemies and Control*, ed. P. Harrewijn & A. Minks. Amsterdam: Elsevier, pp. 289–297.

(1987c). Parthenogenetic reproduction and the rate of increase in aphids. In *Aphids, their Biology, Natural Enemies and Control*, ed. P. Harrewijn & A. Minks. Amsterdam: Elsevier, pp. 269–287.

(1990a). Evolutionary aspects of parthenogenetic reproduction in aphids. *Acta Phytopathology and Entomology Hungarica* **25**, 41–56.

(1990b). Population dynamics and abundance of deciduous tree-dwelling aphids. In *Population Dynamics of Forest Insects*, ed. M. Hunter, N. Kidd, S. R. Leather & A. D. Watt. Andover: Intercept, pp. 11–23.

(1998). *Aphid Ecology*, 2nd edn. London: Chapman & Hall.

(2000). *Insect Predator–Prey Dynamics: Ladybird Beetles and Biological Control.* Cambridge: Cambridge University Press.

(2003). Climate change and phenological asynchrony. *Ecological Entomology* **28**, 380–381.

Dixon, A. F. G., Croghan, P. C. & Gowing, R. P. (1990). The mechanism by which aphids adhere to smooth surfaces. *Journal of Experimental Biology* **152**, 243–253.

Dixon, A. F. G. & Dharma, T. R. (1980). 'Spreading the risk' in developmental mortality: size, fecundity and reproductive rate in the black bean aphid. *Entomologia Experimentalis et Applicata* **28**, 301–312.

Dixon, A. F. G. & Kindlmann, P. (1990). Role of plant abundance in determining the abundance of herbivorous insects. *Oecologia* **83**, 281–283.

(1998a). Generation time ratio and the effectiveness of ladybirds as classical biological control agents. *Proceedings 6th Australasian Applied Entomological Research Conference* **1**, 314–320.

(1998b). Population dynamics of aphids. In *Insect Populations in Theory and in Practice*, ed. J. P. Dempster & I. F. G. McLean. Dordrecht: Kluwer Academic Publishers, pp. 207–230.

Dixon, A. F. G., Kindlmann, P., Leps, J. & Holman, J. (1987). Why are there so few species of aphids, especially in the tropics? *American Naturalist* **129**, 580–592.

Dixon, A. F. G. & Kundu, R. (1997). Trade-off between reproduction and length of adult life in males and mating females of aphids. *European Journal of Entomology* **94**, 105–109.

Dixon, A. F. G. & Logan, M. (1972). Population density and spacing in the sycamore aphid, *Drepanosiphum platanoides* (Schr.), and its relevance to the regulation of population growth. *Journal of Animal Ecology* **41**, 751–759.

(1973). Leaf size and availability of space to the sycamore aphid, *Drepanosiphum platanoides*. *Oikos* **24**, 58–63.

Dixon, A. F. G. & McKay, S. (1970). Aggregation in the sycamore aphid *Drepanosiphum platanoides* (Schr.) (Hemiptera: Aphididae) and its relevance to the regulation of population growth. *Journal of Animal Ecology* **39**, 439–454.

Dixon, A. F. G. & Mercer, D. R. (1983). Fight behaviour in the sycamore aphid: factors affecting take-off. *Entomologia Experimentalis et Applicata* **33**, 43–49.

Dixon, A. F. G. & Russel, R. J. (1972). The effectiveness of *Antocoris nemorum* and *A. confusus* (Hemiptera: Anthocoridae) as predators of the sycamore aphid, *Drepanosiphum platanoides*. II. Searching behaviour and the incidence of predation in the field. *Entomologia Experimentalis et Applicata* **15**, 35–50.

Dixon, A. F. G. & Stewart, W. A. (1975). Function of the siphunculi in aphids with particular reference to the sycamore aphid, *Drepanosiphum platanoides*. *Journal of Zoology (London)* **175**, 279–289.

Dixon, A. F. G., Wellings, P. W., Carter, C. & Nichols, J. F. A. (1993). The role of food quality and competition in shaping the seasonal cycle in the reproductive activity of the sycamore aphid. *Oecologia* **95**, 89–92.

Doebeli, M. & Ruxton, G. D. (1997). Evolution of dispersal rates in metapopulation models: branching and cyclic dynamics in phenotype space. *Evolution* **51**, 1730–1741.

Douglas, A. E. (1989). Mycetocyte symbiosis in insects. *Biological Reviews* **69**, 409–434.

(1995). The ecology of symbiotic micro-organisms. *Advances in Ecological Research* **26**, 69–103.

(2000). Reproductive diapause and the bacterial symbiosis in the sycamore aphid *Drepanosiphum platanoidis*. *Ecological Entomology* **25**, 256–261.

Douglas, A. E. & Dixon, A. F. G. (1987). The mycetocyte symbiosis of aphids variation with age and morph in virginoparae of *Megoura viciae* and *Acyrthosiphon pisum*. *Journal of Insect Physiology* **33**, 109–113.

Eastop, V. F. (1991). Host plant range and virus transmission by aphids. *Fitopatologia Brasiliansis* **16**, 241–245.

Elton, C. S. (1927). *Animal Ecology*. London: Sidgwick & Jackson.

(1966). *The Pattern of Animal Communities*. London: Methuen & Co. Ltd.

Errington, P. L. (1934). Vulnerability of a bobwhite population to predation. *Ecology* **15**, 110–127.

(1946). Predation and vertebrate populations. *Quarterly Review of Biology* **21**, 145–177.

Errington, P. L. & Hamerstrom, F. N. (1936). The northern bobwhite's winter territory. *Research Bulletin Iowa Agricultural Experimental Station* **201**.

Fisher, R. A. (1930). *The genetical theory of natural selection*. Oxford: Oxford University Press.

Fox, R. C. & Griffith, K. H. (1977). Pine seedling growth loss caused by cinarian aphids in South Carolina. *Journal of Georgia Entomological Society* **12**, 13–29.

Fukatsu, T. (2001). Secondary intracellular symbiotic bacteria in aphids of the genus *Yamatocallis* (Homoptera: Aphididae: Drepanosiphinae). *Applied and Environmental Microbiology* **67**, 5315–5320.

Fukatsu, T. & Ishikawa, H. (1992). Soldier and male of an eusocial aphid *Colophina arma* lack endosymbiont: implications for physiological and evolutionary interaction between host and symbiont. *Journal of Insect Physiology* **38**, 1033–1042.

Furuta, K. (1986). Host preference and population dynamics in an autumnal population of the maple aphid, *Periphyllus californiensis* Shinji (Homoptera, Aphididae). *Journal of Applied Entomology* **102**, 93–100.

(1988). Annual alternating population size of the thuja aphid, *Cinara tujafilina* (Del Guercio), and the impact of syrphids and disease. *Journal of Applied Entomology* **105**, 344–354.

(1990a). Early budding of *Acer palmatum* caused by shade; intraspecific heterogeneity of the host for the maple aphid. *Bulletin Tokyo University Forests* **82**, 137–145.

(1990b). Seeding behaviour of *Acer amoenum* and the effect of the infestation of aphids. *Bulletin Tokyo University Forests* **82**, 147–156.

(1994). Influence of the maple aphid, *Periphyllus californiensis*, on the length of long shoots and leaves of young *Acer amoenum*. *Journal Japanese Forestry Society* **76**, 263–269.

(2003). Effects of phenology and natural enemies on long term population dynamics of the maple aphid (*Periphyllus californiensis*) on *Acer palmatum* trees. *Journal of Tree Health* **7**, 7–14.

Furuta, K. & Sakamoto, N. (1984). Seasonal fluctuation of the population density of the maple aphid (*Periphyllus californiensis* Shinji; Hom., Aphididae). *Bulletin Tokyo University Forests* **73**, 97–113.

Furuta, K., Takai, M. & Funatsu, T. (1983). Effects of an infestation of *Cinara bogdanowi ezanoana* Inouye (Hemiptera, Lachnidae) on the growth of *Picea glehnii* Mast. *Journal of Japanese Forestry Society* **65**, 166–171.

Gandon, S. & Michalakis, Y. (1999). Evolutionary stable dispersal rate in a metapopulation with extinctions and kin competition. *Journal of Theoretical Biology* **199**, 275–290.

Gange, A. C. (1985). The ecology of the alder aphid (*Pterocallis alni* (De Geer)) and its role in integrated pest management. PhD Thesis, University of London.

Garthside, S. (1928). Quantitative studies on the insect fauna of Jack Pine environment. PhD Thesis, University of Minnesota.

Gaston, K. J. (1994). *Rarity*. London: Chapman & Hall.

Gaston, K. J. & Lawton, J. H. (1988a). Patterns in the distributions of insect populations. *Nature* **331**, 709–712.

(1988b). Patterns in body size, population dynamics and regional distribution of bracken herbivores. *American Naturalist* **132**, 662–680.

(1990). Effect of scale and habitat on the relationship between regional distribution and abundance. *Oikos* **58**, 329–335.

Gilpin, M. E. (1975). *Group Selection in Predator–Prey Communities*. Prenceton: Princeton University Press.

Glen, D. M. (1971). The role of the black-kneed capsid *Blepharidopterus angulatus* (Fall.) in regulating the numbers of the lime aphid *Eucallipterus tiliae* (L.). PhD Thesis, University of Glasgow.

Godfray, H. C. J. (1994). *Parasitoids; Behavioral and Evolutionary Ecology*. Princeton: Princeton University Press.

Grafen, A. (1990). Biological signals as handicaps. *Journal of Theoretical Biology* **144**, 517–546.

Grier, C. C. & Vogt, D. J. (1990). Effects of aphid honeydew on soil nitrogen availability and net primary production in an *Alnus rubra* plantation in Western Washington. *Oikos* **57**, 114–118.

Gundersen, P., Emmett, B. A., Kjonaas, O. J., Koopmans, C. J. & Tietema, A. (1998). Impact of nitrogen deposition on nitrogen cycling in forests: a synthesis of NITREX data. *Forest Ecology and Management* **101**, 37–55.

Hairston, N. G., Smith, F. E. & Slobodkin, L. B. (1960). Community structure, population control, and competition. *American Naturalist* **94**, 421–425.

Hajek, A. E. & Dahlsten, D. L. (1988). Distribution and dynamics of aphid (Homoptera: Drepanosiphidae) populations on *Betula pendula* in Northern California. *Higardia* **56**, 1–33.

Hamilton, P. A. (1969). The role of hymenopterous parasites in the control of the sycamore aphid. PhD Thesis, University of Glasgow.

Hamilton, P. A. (1973). The biology of *Aphelinus flavus* [Hym. Aphelinidae], a parasite of the sycamore aphid *Drepanosiphum platanoides* [Hemipt. Aphididae]. *Entomophaga* **18**, 449–462.

(1974). The biology of *Monoctonus pseudoplatani, Trixys cirsii* and *Dyscritulus planiceps*, with notes on their effectiveness as parasites of the sycamore aphid, *Drepanosiphum platanoides. Annales de la Société Entomologique de France* **10**, 821–840.

Hamilton, W. D. (1967). Extraordinary sex ratios. *Science* **156**, 477–488.

(1987) Kinship, recognition and disease: constraints of social evolution. In *Animal Societies: Theories and Facts*, ed. Y. Ito, J. L. Brown & J. Kikkawa. Tokyo: Japan Scientific Societies Press, pp. 81–102.

Hamilton, W. D. & Brown, S. P. (2001). Autumn tree colours as a handicap signal. *Proceedings Royal Society London B* **268**, 1489–1493.

Hamilton, W. D. & May, R. M. (1977). Dispersal in stable habitats. *Nature* **269**, 578–581.

Harada, H. & Ishikawa, H. (1993). Gut microbe of aphid closely related to its intracellular symbiont. *Bio Systems* **31**, 185–191.

Harper, J. L. (1977). *Population Biology of Plants*. London: Academic Press.

Harrington, R., Woiwod, I. & Sparks, T. (1999). Climate change and trophic interactions. *Trends in Ecology and Evolution* **14**, 146–150.

Haukioja, E. & Hakala, T. (1975). Herbivore cycles and periodic outbreaks. Formulation of a general hypothesis. *Report Kevo Subarctic Research Station* **12**, 1–9.

Heie, O. E. (1967). Studies on fossil aphids (Homopter: Aphidoidea). *Spolia Zoologica Musei Hauniensis* **26**, 1–274.

Hengeveld, R. & Hacek, J. (1982). The distribution of abundance. 1. Measurements. *Journal of Biogeography* **9**, 303–306.

Hoch, W. A., Zeldin, E. L. & McCowan, B. (2001). Physiological significance of anthocyanins during autumnal leaf senescence. *Tree Physiology* **21**, 1–8.

Hodgson, J. G. (1986). Commonness and rarity in plants, with special reference to the Sheffield flora. Part III. Taxonomic and evolutionary aspects. *Biological Conservation* **36**, 275–296.

Holler, C., Borgemeister, C., Haardt, H. & Powell, W. (1993). The relationship between primary parasitoids and hyperparasitoids of cereal aphids: an analysis of field data. *Journal of Animal Ecology* **62**, 12–21.

Holloway, G. J. (2003). Insect conservation – where are we going? *Antenna* **27**, 320–323.

Holt, R. D. (1997). On the evolutionary stability of sink populations. *Evolutionary Ecology* **11**, 723–731.

Holt, R. D. & Lawton, J. H. (1994). The ecological consequences of shared natural enemies. *Annual Review of Ecology and Systematics* **25**, 495–520.

Holt, R. D. & McPeek, M. A. (1996). Chaotic population dynamics favours the evolution of dispersal. *American Naturalist* **148**, 709–718.

Hopkins, G. W. (1996). Rarity in tree aphids. PhD thesis, University of East Anglia.

Hopkins, G. W. & Dixon, A. F. G. (1997). Enemy-free space and the feeding niche of an aphid. *Ecological Entomology* **22**, 271–274.

Hopkins, G. W., Thacker, J. I., Dixon, A. F. G., Waring, P. & Telfer, M. G. (2002). Identifying rarity in insects: the importance of host plant range. *Biological Conservation* **105**, 293–307.

Huffaker, C. B. (1957). Fundamentals of biological control of weeds. *Hilgardia* **27**, 101–157.

Huffaker, C. B., Berryman, A. & Turchin, P. (1999). Dynamics and regulation of insect populations. In *Ecological Entomology*, ed. C. B. Huffaker & A. P. Gutierrez. London: Academic Press, pp. 269–312.

Hussey, N. W. (1952). A contribution to the bionomics of the green spruce aphid (*Neomyzaphis abietina* Walker). *Scottish Forestry* **6**, 121–130.

Huxley, T. H. (1858). On the agamic reproduction and morphology of *Aphis*: Part 1. *Transactions of the Linnean Society* **22**, 193–219.

Ito, Y. (1994). A new epoch in joint studies of social evolution: molecular and behavioural ecology of aphid soldiers. *Trends in Ecology and Evolution* **9**, 363–365.

Jackson, J. (1970). Vertical migration of the sycamore aphid *Drepanosiphum platanoides* (Schr.). PhD Thesis, University of Glasgow.

Janzen, D. H. (1977). What are dandelions and aphids? *American Naturalist* **111**, 586–589.

Jefferies, J. J. & Lawton, J. H. (1984). Enemy-free space and the structure of ecological communities. *Biological Journal of the Linnean Society* **23**, 269–286.

Johansson, A. S. (1958). Relation of nutrition to endocrine-reproductive functions in the milkweed bug *Oncopeltus fasciatus* (Dallas) (Heteroptera: Lygaeidae). *Nytt Magasin for Zoologi* **7**, 1–132.

Juronis, V. (2001). *Eucallipterus tiliae* L. – a parasite of street plantings in Lithuanian cities. *Aphids and other Homopterus Insects* **8**, 131–134.

Kaakeh, W., Pfeiffer, D. G. & Marini, R. P. (1992a). Combined effects of spirea aphid (Homoptera: Aphididae) and nitrogen fertilization on shoot growth, dry matter accumulation, and carbohydrate concentration in young apple trees. *Journal of Economic Entomology* **85**, 496–506.

(1992b). Combined effect of spirea aphid (Homoptera: Aphididae) and nitrogen fertilization on net photosynthesis, total chlorophyll content, and greenness of apple leaves. *Journal of Economic Entomology* **85**, 939–946.

Kennedy, C. E. J. (1986). Attachment may be a basis for specialization in oak aphids. *Ecological Entomology* **11**, 291–300.

Kennedy, J. S. & Crawley, L. (1967). Spaced-out gregariousness in sycamore aphids *Drepanosiphum platanoides* (Schrank) (Hemiptera, Callaphididae). *Journal of Animal Ecology* **36**, 147–170.

Kidd, N. A. C. (1975). The behavioural interaction of the lime aphid (*Eucallipterus tiliae* (L.)) and their role in regulating population numbers. PhD Thesis, University of Glasgow.

(1990a). Population dynamics of the large pine aphid, *Cinara pinea* (Mordv.). I. Simulation of laboratory populations. *Researches on population Ecology* **32**, 189–208.

(1990b). Population dynamics of the large pine aphid, *Cinara pinea* (Mordv.). II. Simulation of field populations. *Researches on Population Ecology* **32**, 209–226.

(1990c). A synoptic model to explain long-term population changes in the large pine aphid. In *Population Dynamics of Forest Insects*, ed. A. D. Watt, S. R. Leather, M. D. Hunter & N. A. C. Kidd, Andover: Intercept, pp. 317–327.

Kieffer, J.-J. (1896). Observations sur les Diplosis, et diagnoses de cinq espèces nouvelles [Dipt.]. *Bulletin de la Société Entomologique de France* **65**, 382–384.

Kindlmann, P. & Dixon, A. F. G. (1989). Developmental constraints in the evolution of reproductive strategies: telescoping of generations in parthenogenetic aphids. *Functional Ecology* **3**, 531–537.

(1993). Optimal foraging in ladybird beetles (Coleoptera: Coccinellidae) and its consequences for their use in biological control. *European Journal of Entomology* **90**, 443–450.

Kindlmann, P., Dixon, A. F. G. & Brough, C. N. (1992). Intra- and interspecific relationships of reproductive investment to body weight in aphids. *Oikos* **64**, 548–552.

Knabe, S. (1999). The ecology of the subspecies of the pea aphid. PhD Thesis, University of East Anglia.

Komazaki, S. (1986). The inheritance of egg hatching time of the spirea aphid, *Aphis citricola* van der Goot (Homoptera, Aphididae) on two winter hosts. *Kontyû* **54**, 48–53.

Komatsu, T. & Akimoto, S. (1995). Genetic differentiation as a result of adaptation to the phenologies of individual host trees in the galling aphid *Kaltenbachiella japonica*. *Ecological Entomology* **20**, 33–42.

Krebs, J. R. & Davies, N. B. (1987). *An Introduction to Behavioural Ecology*. Oxford: Blackwells.

Lawton, J. H. (1986). The effect of parasitoids on phytophagous insect communities. In *Insect Parasitoids*, ed. J. K. Waage & D. Greathead, London: Academic Press, pp. 265–287.

(1989). What is the relationship between population density and body size in animals? *Oikos* **55**, 429–434.

Leather, S. R. (1996). Colonisation and distribution patterns of sycamore aphid on sycamore trees in south-east Britain. *Bulletin British Ecological Society* **27**, 214–218.

(2000). Herbivory, phenology, morphology and the expression of sex in trees: who is in the driver's seat? *Oikos* **90**, 194–196.

Leckstein, P. M. & Llewellyn, M. (1973). Effect of dietary amino acids on the size and alary polymorphism of *Aphis fabae*. *Journal of Insect Physiology* **19**, 973–980.

Lees, A. D. (1960). The role of photoperiod and temperature in the determination of parthenogenetic and sexual forms in the aphid *Megoura viciae* Buckton. II. The operation of the 'interval timer' in young clones. *Journal of Insect Physiology* **4**, 154–175.

(1966). The control of polymorphism in aphids. *Advances in Insect Physiology* **3**, 207–277.

Lees, A. D. & Hardie, J. (1988). The organs of adhesion in the aphid *Megoura viciae*. *Journal of Experimental Biology* **136**, 201–208.

Leopold, A. (1943). Deer irruptions. *Wisconsin Conservation Bulletin* **8**, 4–11.

Levin, R. (1968). *Evolution in Changing Environments*. Princeton: Princeton University Press.

Liao, H. T. & Harris, M. K. (1985). Population growth of the black-margined aphid on pecan in the field. *Agricultural Ecosystems and Environment* **12**, 253–261.

Llewellyn, M. J. (1970). The ecological energetics of the lime aphid (*Eucallipterus tiliae* L.) and its effect on tree growth. PhD Thesis, University of Glasgow.

(1972). The effect of the lime aphid, *Eucallipterus tiliae* L. (Aphididae) on the growth of lime *Tilia* x *vulgaris* Hayne. I. Energy requirements of the aphid populations. *Journal of Applied Ecology* **9**, 261–282.

(1975). The effects of the lime aphid (*Eucallipterus tiliae* L.) (Aphididae) on the growth of lime *Tilia* x *vulgaris* Hayne. II. The primary production of saplings and mature trees, the energy drain imposed by the aphid populations and revised standard deviations of aphid population energy budgets. *Journal of Applied Ecology* **12**, 15–23.

Lorriman, F. (1980). The ecology and biology of the oak aphid *Tuberculatus (Tuberculoides) annulatus* Hartig. PhD Thesis, University of London.

Lotka, A. J. (1924). *Elements of Physical Biology*. Baltimore: Williams & Wilkins.

Loxdale, H. D., Hardie, J., Halbert, S., Foottit, R., Kidd, N. A. C. & Carter, C. I. (1993). The relative importance of short- and long-range movement in flying aphids. *Biological Reviews* **68**, 291–311.

MacArthur, R. & Wilson, E. O. (1967). *Theory of Island Biogeography*. Princeton: Princeton University Press.

Mackauer, M. & Völkl, W. (1993). Regulation of aphid populations by aphidiid wasps: does parasitoid foraging behaviour or hyperparasitism limit impact? *Oecologia* **94**, 339–350.

Macfayden, A. (1964). Energy flow in ecosystems and its exploitation by grazing. In *Grazing in Terrestrial and Marine Environments*, ed. D. J. Crisp. Oxford: Blackwells, pp. 3–20.

Maquelin, C. (1974). Observations sur la biologie et l'ecologie d'un puceron utile a l'apiculture: *Buchneria pectinatae* (Nördl.)(Homoptera, Lachnidae). PhD Thesis, L'Ecole Polytechnique Federale De Zurich.

Martins, E. P. (1996). Conducting phylogenetic comparative studies when the phylogeny is not known. *Evolution* **50**, 12–22.

Maynard-Smith, J. (1976). Group selection. *Quarterly Review of Biology* **51**, 277–283.

McKinney, M. L. (1999). High rates of extinction and threat in poorly studied taxa. *Conservation Biology* **13**, 1273–1281.

McNaughton, F. C. (1970). The energy required by sycamore aphids and their effect on the growth of sycamore. PhD Thesis, University of Glasgow.

McPeek, M. A. & Holt, R. D. (1992). The evolution of dispersal in spatially and termporally varying environments. *American Naturalist* **140**, 1010–1027.

Mercer, D. R. (1979). Flight behaviour of the sycamore aphid *Drepanosiphum platanoidis* Schr. PhD Thesis, University of East Anglia.

Milne, A. (1957a). Theories of natural control of insect populations. *Cold Spring Harbour Symposium Quantitative Biology* **22**, 253–271.

(1957b). The natural control of insect populations. *Canadian Entomologist* **89**, 193–213.

Mittler, T. E. (1957). Studies on the feeding and nutrition of *Tuberolachnus salignus* (Gmelin) (Homopter, Aphididae). I. The uptake of phloem sap. *Journal of Experimental Biology* **34**, 334–341.

(1958). Studies on the nutrition of *Tuberolachnus salignis* (Gmelin) (Homoptera, Aphididae). III. The nitrogen economy. *Journal of Experimental Biology* **35**, 626–638.

Mittler, T. E. & Wipperfurth, T. (1988). Hatching and diapause development of the eggs from crosses between Biotypes C and E of the aphid *Schizaphis graminum* (Homoptera: Aphididae). *Entomologia Generalis* **13**, 247–249.

Moran, N. A. & Baumann, P. (1994). Phylogenetics of cytoplasmically inherited microorganism of arthropods. *Trends in Ecology and Evolution* **9**, 15–20.

Mordvilko, A. K. (1908). Beitrage zur Biologie der Pfanzenläuse, Aphididae Passerini. *Biologische Zentrablatt* **28**, 631–639.

Morren, C. H. (1836). Mémoire sur l'emigration de puceron du pêcher (*Aphis persicae*), et sur les caractères et l'anatomie de cette espèce. *Annales des Sciences Naturelles* **6**(2), 65–93.

Müller, C. B. & Godfray, H. C. J. (1997). Apparent competition between two aphid species. *Journal of Animal Ecology* **66**, 57–64.

Murray, M. B., Cannell, M. G. R. & Smith, R. I. (1989). Date of budburst of fifteen tree species in Britain following climatic warming. *Journal of Applied Ecology* **26**, 693–700.

Nee, S., Read, A. F., Greenwood, J. J. D. & Harvey, P. H. (1991). The relationship between abundance and body size in British birds. *Nature* **351**, 312–313.

Nicholson, A. J. (1933). The balance of animal populations. *Journal of Animal Ecology* **2** (Suppl. 1), 132–178.

(1954). An outline of the dynamics of animal populations. *Australian Journal of Zoology* **2**, 9–65.

Odum, E. P. (1953). *Fundamentals of Ecology*. Philadelphia: Saunders.

Olkowski, W. (1973). A model ecosystem management program for street tree insects in Berkeley, California. PhD Thesis, University of California, Berkeley.

Olkowski, W., Olkowski, H. & van den Bosch, R. (1982). Linden aphid parasite establishment. *Environmental Entomology* **11**, 1023–1025.

Owen, D. F. (1978). Why do aphids synthesize melezitose? *Oikos* **31**, 264–267.

(1980a). How plants may benefit from the animals that eat them? *Oikos* **35**, 230–235.

(1980b). Response to Petelle's comments. *Oikos* **35**, 128.

Owen, D. F. & Wiegert, R. G. (1976). Do consumers maximise plant fitness? *Oikos* **27**, 488–492.

(1981). Mutualism between grasses and grazers: an evolutionary hypothesis. *Oikos* **36**, 376–378.

Owen, R. (1849). *On Parthenogenesis or the Successive Production of Procreating Individuals from a Single Ovum*. London: John van Voorst.

Pearl, R. & Reed, L. J. (1920). On the rate of growth of the population of the United States since 1790 and its mathematical representation. *Proceedings National Academy Science, USA* **6**, 275–288.

Pennacchio, F., Digilio, M. C. & Tremblay, E. (1995). Biochemical and metabolic alterations in *Acyrthosiphon pisum* parasitized by *Aphidius ervi*. *Archives of Insect Biochemistry and Physiology* **30**, 351–367.

Petelle, M. (1980). Aphids and melezitose: a test of Owen's 1978 hypothesis. *Oikos* **35**, 127–128.

Pitman, N. C. A. & Jorgensen, P. M. (2002). Estimating the size of the world's threatened flora. *Science* **298**, 989–999.

Plantegenest, M. & Kindlmann, P. (1999). Evolutionarily stable strategies of migration in heterogenous environments. *Evolutionary Ecology* **13**, 229–244.

Ponsen, M. B. (1972). The site of potato leafroll virus multiplication in its vector, *Myzus persicae*. *Wageningen Agricultural University Papers* No. 72-16.

(1991). Structure of the digestive system of aphids, in particular *Hyalopterus* and *Coloradoa*, and its bearing on the evolution of filter chambers in the Aphidoidea. *Wageningen Agricultural University Papers* No. 91-5.

Rabinowitz, D. (1981). Seven forms of rarity. In *The Biological Aspects of Rare Plant Conservation*, ed. H. Synge. New York: Wiley, pp. 205–217.

Rabinowitz, D., Rapp, J. K. & Dixon, P. M. (1984). Competitive abilities of grass species: means of persistence or cause of abundance. *Ecology* **65**, 1144–1154.

Rana, J. S., Dixon, A. F. G. & Jarosik, V. (2002). Costs and benefits of prey specialization in a generalist insect predator. *Journal of Animal Ecology* **71**, 15–22.

Réaumur, R. P. de (1737). *Mémoires pour servir à l'histoire des insectes.* **111**(9) Paris: De L'imprimiere Royale, pp. 332–350.

Renshaw, E. (1991). *Modelling Biological Populations in Space and Time.* Cambridge: Cambridge University Press.

Rhabé, Y., Digilio, M. C., Febvay, G., Guillaud, J., Fanti, P. & Pennacchio, F. (2002). Metabolic and symbiotic interactions in amino acid pools of the pea aphid, *Acyrthosiphon pisum*, parasitized by the braconid *Aphidius ervi. Journal of Insect Physiology* **48**, 507–516.

Ricklefs, R. E. (1990). *Ecology*, 3rd edn. New York: Freeman & Co.

Ricklefs, R. E. & Latham, R. E. (1992). Intercontinental correlation of geographical ranges suggests stasis in ecological traits of relict genera of temperate perennial herbs. *American Naturalist* **139**, 1305–1321.

Robert, Y. & Rouzé-Jouan, J. (1976). Activité saisonnière de vol des pucerons [*Hom. Aphididae*] dans l'ouest de la France. Résultats de neuf années de piégeage (1967–1975). *Annales de la Société Entomologique de France (NS)* **12**, 671–690.

Root, R. B. (1967). The niche exploitation pattern of the blue-gray gnatcatcher. *Ecological Monographs* **37**, 317–350.

Russel, R. J. (1968). Certain aspects of the ecology of *Anthocoris nemorum* (L.) and *Anthocoris confusus* Reuter (*Hemiptera: Anthocoridae*). PhD Thesis, University of Glasgow.

Satoo, T. (1970). A synthesis of studies by the harvest method. *Ecological Studies* **1**, 55–72.

Scheurer, S. (1964). Untersuchungen zum Massenwechsel einiger Fichten bewohnender Lachnidenarten im Harz. *Biologisches Zentrablatt* **83**, 427–467.

    (1971). Biologische und ökologische Beobachtungen an auf *Pinus* lebenden Cinarinen im Bereich der Dübener Heide (DDR) während de Jahre 1965–1967. *Hercynia (Leipzig)* **8**, 108–144.

Schulze, E.-D, (2000). The carbon and nitrogen cycle of forest ecosystems. In *Carbon and Nitrogen Cycling in European Forest Ecosystems*, ed. E.-D. Schulze, vol. 142. Berlin: Springer Verlag, pp. 3–13.

Seger, J. (1983). Partial bivoltinism may cause alternating sex-ratio biases that favour eusociality. *Nature* **301**, 59–62.

Sequeira, R. & Dixon, A. F. G. (1997). Population dynamics of tree dwelling aphids: the importance of seasonality and time scale. *Ecology* **78**, 2603–2610.

Shearer, J. W. (1976). Polymorphism and population ecology of the European maple aphid, *Periphyllus testudinaceus* (Fernie). PhD Thesis, University of Glasgow.

Shirt, D. B. (1987). *British Red Data Books: 2. Insects.* Peterborough: Nature Conservancy Council.

Sibbett, G. S., Bettiga, L. & Bailey, M. (1982). Walnut aphid becoming a costly midsummer pest. *California Agriculture* **36**, 21–22.

Slansky, F. & Feeny, P. (1977). Stabilisation of the rate of nitrogen accumulation by larvae of the cabbage butterfly on wild and cultivated food plants. *Ecological Monographs* **47**, 209–228.

Smalley, A. E. (1960). Energy flow of a salt marsh grasshopper population. *Ecology* **41**, 672–677.

Smith, H. S. (1935). The role of biotic factors in the determination of population densities. *Journal of Economic Entomology* **28**, 873–898.

Smith, J. D. (1948). Symbiotic micro-organisms of aphids and fixation of atmospheric nitrogen. *Nature* **162**, 930–931.

Smith, J. S. & Tedders, W. L. (1980). Light measurements for studying sooty mold growth on simulated pecan foliage. *Transaction of the American Society of Agricultural Engineers* **23**, 481–484.

Southwood, T. R. E. (1976). Bionomic strategies and population parameters. In *Theoretical Ecology*, ed. R. M. May. Oxford: Blackwell Scientific Publications, pp. 26–48.

Stadler, B., Michalzik, B. & Müller, T. (1998). Linking aphid ecology with nutrient fluxes in coniferous forests. *Ecology* **79**, 1514–1525.

Stadler, B. & Michalzik, B. (2004). Phyllosphere ecology in a changing environment: the role of insects in forested ecosystems. In *Biogeochemistry of Forested Catchments in a Changing Environment: a case study in NE Bavaria*, ed. E. Matzner. *Ecological Studies* **172**, 251–270.

Stork, N. E. & Lyal, C. H. C. (1993). Extinction or co-extinction rates. *Nature* **366**, 307–307.

Straw, N. A., Halldórsson, G. & Benedikz, T. (1998). Damage sustained by individual trees: empirical studies on the impact of the green spruce aphid. In *The Green Spruce Aphid in Western Europe: Ecology, Staus, Impacts and Prospects for Management*, ed. K. R. Day, G. Halldórsson, S. Harding & N. A. Straw, Forestry Commission Technical Paper 24, pp. 15–31.

Stroyan, H. L. G. (1977). *Homoptera Aphidoidea Chaitophoridae and Callaphididae. Handbook for the Identification of British Insects*, vol. 11, part 4(a). London: Royal Entomological Society of London.

Tang, Y. Q., Yokomi, R. K. & Gagné, R. J. (1994). Life history and description of *Endaphis maculans* (Diptera: Cecidomyiidae), an endoparasitoid of aphids in Florida and the Caribbean basin. *Annals Entomological Society of America* **87**, 523–531.

Taylor, L. R. (1974a). Insect migration, flight periodicity and boundary layer. *Journal of Animal Ecology* **43**, 225–238.

(1974b). Monitoring change in the distribution and abundance of insects. *Report Rothamsted Experimental Station 1973* **2**, 202–239.

Tedders, W. (1978). Important biological and morphological characteristics of the foliar-feeding aphids of pecan. *USDA Technical Bulletin* No. 1579.

Tedders, W. L. & Smith, J. S. (1976). Shading effect on pecan by sooty mold growth. *Journal of Economic Entomology* **69**, 551–553.

Tedders, W. L., Smith, J. S. & White, A. W. (1981a). Experiment to determine the effect of feeding by *Monellia caryella* (Fitch) and of simulated honeydew on pecan seedlings in the greenhouse. *Journal of Georgia Entomological Society* **16**, 515–517.

Tedders, W. L., Wood, B. W. & Snow, J. W. (1981b). Effects of feeding by *Monelliopsis nigropunctata*, *Monellia caryella*, and *Melanocallis caryaefoliae* on growth of pecan seedlings in the greenhouse. *Journal of Economic Entomology* **75**, 287–291.

Tedders, W. L. & Wood, B. W. (1985). Estimate of the influence of feeding by *Monelliopsis pecanis* and *Monellia caryella* (Homoptera: Aphididae) on the fruit, foliage, carbohydrate reserves, and tree productivity of mature 'Stuart' pecans. *Journal of Economic Entomology* **78**, 642–646.

Thacker, J. I., Hopkins, G. W. & Dixon, A. F. G. (2003). The co-extinction threat to insect herbivores: aphids and coccids on endangered trees.

Tjallingii, W. F. (1978). Mechanoreceptors of the aphid labium. *Entomologia Experimentalis et Applicata* **24**, 531–537.

Tóth, L. (1937). Entwicklungszyklus und Symbiose von *Pemphugus spirothecae* Pass. (Aphidina). *Zeitschrift für Morphologie und Okologie der Tiere* **33**, 412–437.

Trager, W. (1970). *Symbiosis*. New York: Van Nostran.

Tremblay, E. & Iaccarino, F. M. (1971). Notizie sull' ultra strultura dei trofociti di *Aphidius matricariae* Haliday. *Bolletino del Laboratorio di Entomologia Agraria 'Filippo Silvestri'* **29**, 305–314.

Tremblay, E. & Ponzi, R. (1999). Ultrastructural observation on symbiont degeneration in the male line of *Pseudaulacaspis pentagona* (Targioni Tozzeti) (Hemiptera: Coccoidea: Diaspididae). *Entomologica* **33**, 157–163.

Tsai, C. S., Killham, K. & Cresser, M. S. (1997). Dynamic response of microbial biomass, respiration rate and ATP to glucose additions. *Soil Biology and Biochemistry* **29**, 1249–1256.

Turchin, P. (1990). Rarity of density dependence or population regulation with lags? *Nature* **344**, 660–663.

(2001). Does population ecology have general rules? *Oikos* **94**, 17–26.

Turchin, P. & Taylor, A. D. (1992). Complex dynamics in ecological time series. *Ecology* **73**, 289–305.

Uichanco, L. B. (1924). Studies on the embryology and postnatal development of the Aphididae with special reference to the history of the 'symbiotic organ' or 'mycetome'. *Philippine Journal of Science* **24**, 143–247.

Varley, G. C., Gradwell, G. R. & Hassell, M. P. (1973). *Insect Population Ecology: An Analytical Approach*. Oxford: Blackwell Scientific Publications.

Varn, M. W. & Pfeiffer, D. G. (1989). The effect of rosy apple aphid and spirea aphid (Homoptera: Aphididae) on dry matter accumulation and carbohydrate concentration in young apple trees. *Journal of Economic Entomology* **82**, 565–569.

Verhulst, P. F. (1838). Notices sur la loi que la population suit dans son croissement. *Correspondance Mathématique et Physique* **10**, 113–121.

Völkl, W. & Mackauer, M. (1996). 'Sacking' the host: oviposition behaviour of a parasitoid wasp, *Dyscritulus planiceps* (Hymenoptera: Aphididae). *Journal of Insect Behaviour* **9**, 975–980.

Voûte, A. D. (1957). Regulierung der Bevolkerungsdichte von schadlichen Insekten auf geringer Höhe durch Nährplanze (*Myelophilus piniperda* L., *Retina buoliana* Schiff., *Diprion sertifer* Geoffr.). *Zeitschrift für Angewandte Entomologie* **41**, 172–178.

Wade, F. A. (1999). Population dynamics of the sycamore aphid (*Drepanosiphum platanoidis* Schrank). PhD Thesis, Imperial College, University of London.

Walters, K. F. A., Dixon, A. F. G. & Eagles, G. (1984). Non-feeding by adult gynoparae of *Rhopalosiphum padi* and its bearing on the limiting resource in the production of sexual females in host alternating aphids. *Entomologia Experimentalis et Applicata* **36**, 9–12.

Ward, S. A., Leather, S. R. & Dixon, A. F. G. (1984). Temperature prediction and the timing of sex in aphids. *Oecologia* **62**, 230–233.

Ward, S. A., Leather, S. R., Pickup, J. & Harrington, R. (1998). Mortality during dispersal and the cost of host-specificity in parasites: how many aphids find hosts? *Journal of Animal Ecology* **67**, 763–773.

Warrington, S. & Whittaker, J. B. (1985). An experimental field study of different levels of insect herbivory induced by *Formica rufa* predation on sycamore (*Acer pseudoplatanus*) II. Aphidoidea. *Journal of Applied Ecology* **22**, 787–796.

Watt, A. D. & McFarlane, A. M. (2002). Will climate change have a different impact on different trophic levels? Phenological development of winter moth *Operophtera brumata* and its host plants. *Ecological Entomology* **27**, 254–256.

Watt, A. D. & Woiwod, I. P. (1999). The effect of phenological asynchrony on population dynamics: analysis of fluctuations of British macrolepidoptera. *Oikos* **87**, 411–416.

Way, M. J. & Banks, C. J. (1964). Natural mortality of eggs of the black bean aphid, *Aphis fabae* Scop., on the spindle tree, *Euonymus europea* L. *Annals of Applied Biology* **54**, 255–267.

(1967). Intra-specific mechanisms in relation to the natural regulation of numbers of *Aphis fabae* Scop. *Annals of Applied Biology* **59**, 189–205.

Wellings, P. W. (1980). Qualitative changes in the regulation of sycamore aphid numbers. PhD Thesis, University of East Anglia.

(1981). The effect of temperature on the growth and reproduction of two closely related aphid species on sycamore. *Ecological Entomology* **6**, 209–214.

Wellings, P. W., Chambers, R. J., Dixon, A. F. G. & Aikman, D. (1985). Sycamore aphid numbers and population density. I. Some patterns. *Journal of Animal Ecology* **54**, 411–424.

Wellings, P. W. & Dixon, A. F. G. (1987). Sycamore aphid numbers and population density. III. The role of aphid-induced changes in plant quality. *Journal of Animal Ecology* **56**, 161–170.

Wellings, P. W., Leather, S. R. & Dixon, A. F. G. (1980). Seasonal variation in reproductive potential: a programmed feature of aphid life cycles. *Journal of Animal Ecology* **49**, 975–985.

Werren, J. H. & Charnov, E. L. (1978). Facultative sex ratio and population dynamics. *Nature* **272**, 349–350.

White, G. (1887). *The Natural History of Selborne*. London: Walter Scott.

White, P. L. (1970). The effect of aphids on tree growth. PhD Thesis, University of Glasgow.

Wilkaniec, B. (1990). The study of direct noxiousness of the rosy apple aphid (*Dysaphis plantaginea* (Pass.)) on apple (*Malus sp.*). *Annals Academy of Agriculture, Poznan* No. 195.

Wilkinson, D. M., Sherratt, T. N., Phillips, D. M., Wratten, S. D., Dixon, A. F. G. & Young, A. J. (2002). The adaptive significance of autumn leaf colours. *Oikos* **99**, 402–407.

Williams, C. B. (1964). *Patterns in the Balance of Nature*. London: Academic Press.

Wood, B. W. & Tedders, W. L. (1986). Reduced net photosynthesis of leaves from mature pecan trees by three species of pecan aphid. *Journal of Entomological Society* **21**, 355–360.

Wood, B. W., Tedders, W. L. & Dutcher, J. D. (1987). Energy drain by three pecan aphid species (Homoptera: Aphididae) and their influence on in-shell pecan production. *Environmental Entomology* **16**, 1045–1056.

Wood, B. W., Tedders, W. L. & Reilly, C. C. (1988). Sooty mold fungus on pecan foliage suppresses light penetration and net photosynthesis. *HortScience* **23**, 851–853.

Wood, B. W., Tedders, W. L. & Thompson, J. M. (1985). Feeding influence of three pecan aphid species on carbon exchange and phloem integrity of seedling pecan foliage. *Journal of American Horticultural Society* **110**, 393–397.

Wratten, S. D. (1971). The role of the predatory coccinellid *Adalia bipunctata* L., in regulating the numbers of the lime aphid, *Eucallipterus tiliae* L. PhD Thesis, University of Glasgow.

Wynne, I. R., Howard, J. J., Loxdale, H. D. & Brookes, C. P. (1994). Population genetic structure during aestivation in the sycamore aphid *Drepanosiphum platanoidis* (Hemoptera: Drepanosiphididae). *European Journal of Entomology* **91**, 375–383.

Yamaguchi, H. (1976). Biological studies on the todo-fir aphid *Cinara todocola* Inouye, with special reference to its population dynamics and morph determination. *Bulletin of the Government Forest Experimental Station (Japan)* No. 283.

Yamaguchi, H. & Takai, M. (1977). An integrated control system for the todo-fir aphid, *Cinara todocola* Inouye in young todo-fir plantations. *Bulletin of the Government Forest Experiment Station (Japan)* No. 295, 61–96.

Zahavi, A. (1975). Mate selection: a selection for handicap. *Journal of Theoretical Biology* **53**, 205–214.

Zuparko, R. L. (1983). Biological control of *Eucallipterus tiliae* [Hom.: Aphididae] in San Jose, California, through establishment of *Trioxys curvicaudus* [Hym.: Aphidiidae]. *Entomophaga* **28**, 325–330.

Species index

Note: page numbers in *italics* refer to figures and tables

Subject index

Note: page numbers in *italics* refer to figures and tables

telescoping of generations *10*,
  8–11
  rate of increase 10
throughfall 152
traps, insect *33*, 31–3, 37–8,
  130
trees 18
  between-year population dynamics
    105–9
  canopy area 19
  carrying capacity 54–63
  competition 157
  deciduous 18–20
  defence 147–8
    autumn colours 148
  distribution on 29, *30*, 81–2
  fitness 147–61
  geometry 19
  growth *156*, 153–7, *158*
  isolated 118
  mutualism with aphids 151–3
  number of leaves 19, *20*

phenology 135
  aphid fitness 133
recruitment 160–1
resistance 147
  to viral infections 157
root mass 156
seeding 157
seedling establishment 157
single-species stands 118
size 134
virus transmission by aphids 157
Turkey oak aphid system 69–70
Turkey oak aphid 6, 140
  population dynamics 93–111

umbrella species 162

virus transmission 157
viviparity 6–8

wind 40–2
  perturbations *41*, 40–2

Printed in the United States
By Bookmasters